Minerals and Rocks 14

W0245722

Editor in Chief
P. J. Wyllie, Chicago, IL

Editors
A. El Goresy, Heidelberg
W. von Engelhardt, Tübingen · T. Hahn, Aachen

Alok K. Gupta   Kenzo Yagi

Petrology and Genesis of Leucite-Bearing Rocks

With 99 Figures and 43 Tables

Springer-Verlag
Berlin Heidelberg New York 1980

Dr. ALOK K. GUPTA
Department of Geology and Geophysics
University of Roorkee
Roorkee 247672, U.P., India

Professor Dr. KENZO YAGI
Department of Geology and Mineralogy
Hokkaido University
Sapporo, Japan

Volumes 1 to 9 in this series appeared under the title
Minerals, Rocks and Inorganic Materials

ISBN-13:978-3-642-67552-2   e-ISBN-13:978-3-642-67550-8
DOI: 10.1007/978-3-642-67550-8

Library of Congress Cataloging in Publication Data. Gupta, Alok K, 1942–. Petrology
and genesis of leucite-bearing rocks. (Minerals and rocks; v. 14). Includes bibliographi-
cal references and index. 1. Leucite. 2. Petrology. I. Yagi, Kenzo, 1914– joint author. II.
Title. III. Series. QE462.L48G86   549′.68   79-27624.

2132/3130-543210

Preface

Many interesting and perplexing questions arise in connection with the highly potassic volcanic association dominated by mafic and ultramafic rocks containing leucite. Its occurrence is very restricted as compared with the olivine-basalt trachyte kindred, but it is distributed at widely scattered points on all the continents, and its chemical and petrographic individuality is both remarkable and constant. A considerable literature is available related to the mineralogy, petrology, geochemistry, phase chemistry, distribution, and origin of this interesting suite of rocks. It seemed that there was a genuine need for a review-synthesis of all these data, which would be intelligible to a wide spectrum of advanced students and professionals in the earth sciences.

The monograph may be divided into two parts. The first part consists of six chapters in which the mineralogical and chemical peculiarities of leucite-bearing rocks and their nomenclature, petrology, mineralogy, distribution, and physical and chemical conditions of formation are discussed. Phase equilibria studies on many leucite-bearing ternary, pseudoternary, quaternary, and pseudoquaternary joins and systems, studied by different investigators at variable temperatures in air, are described in the second part in Chapters 7 to 12. Survival of leucite and formation of pseudoleucite is discussed in Chapter 13. Leucite-bearing synthetic and natural rock systems studied at different temperatures under variable pressures in presence or absence of water, are summarized in Chapters 14 and 15. Chemical affinity between leucite-bearing rocks and kimberlites and some experimental investigations to determine the possible genetic connection between these two groups of rocks are described in Chapter 16. Structural and tectonic control of alkali volcanism with special reference to leucite-bearing lavas is discussed in Chapter 17. In the final chapter a critical review of various hypotheses on the origin of leucite-bearing rocks has been presented. In this chapter, there is an attempt to utilize the knowledge of trace element geochemistry, field and experimental data to elucidate the origin of leucite-bearing magma. Explanation for the abbreviations used in figures and the text is given in the List of Abbreviations. Many figures were obtained directly from the papers of various workers, hence in some cases two slightly different symbols have been used to denote the same phase in two different figures. The second symbol has been shown within brackets in the appendix.

We are especially indebted to Dr. R.H. Grapes, Dr. A. Cundari, Dr. R.W. Le Maitre, and Dr. P.J. Wyllie, who read the entire manuscript and helped us to improve it. We are also grateful to Dr. H.S. Yoder, Dr. A.L. Boettcher, Dr. J. Gittins, Dr. M.J. Le Bas, and Dr. K. Aoki, who read specific chapters and offered constructive comments. However, we take sole responsibility for any errors or omissions. While reading the Chapters 8, 9, and 13, it will be evident that their contents are based on three papers, that one of us (A.K.G.) wrote with Dr. E.G. Lidiak, Dr. A.D. Edgar, and Dr. W.S. Fyfe. We are indebted to these authors for their valuable contribution. We thank Dr. K. Onuma and Dr. Y. Hariya for their help in some high-temperature and high-pressure experiments. Thanks are also due to Dr. A.K. Dubey, Mr. R. Sreedhar, and Mr. G.K. Bhatt for editing the manuscript. We are grateful to Mrs. C. Gupta, who typed a significant portion of the manuscript. We gratefully acknowledge the help of Mr. S. Terada for his care in drafting many diagrams.

One of us (A.K.G.) wishes to dedicate his contribution of the manuscript to his father Mr. K.P. Gupta.

It is a pleasure to acknowledge the friendly relationship which we have enjoyed with our patient editor, Professor P.J. Wyllie. We set and passed many deadlines without once being taken to task for the delays in completing the manuscript.

Above all, we will be very happy if this volume is useful to advanced students of the earth sciences everywhere.

January 1980   ALOK K. GUPTA
                                                      KENZO YAGI

Contents

List of Abbreviations

The abbreviations have been used in text and figures

Lc	Leucite	Sp	Spinel
Fe-Lc	Iron leucite	Wo	Wollastonite
Ne	Nepheline	Ra	Rankinite
Cg	Carnegieite	Mer	Merwinite
Anl	Analcite	Mo	Monticellite
Ks	Kalsilite	La	Larnite
Ab	Albite	Ph	Phlogopite
An	Anorthite	(Phlogo)	
Pl	Plagioclase	Amph	Amphibole
Qtz (Q)	Quartz	Hb	Hornblende
Tr	Tridymite	Ru	Rutile
Or	Orthoclase	Sph	Sphene
K-F	Potash feld-	Il	Ilmenite
	spar	Ap	Apatite
Di	Diopside	Mt	Magnetite
Cpx	Clinopyroxene	L (Liq)	Liquid
En	Enstatite	V	Vapor
Pen	Protoenstatite	Xtl	Crystal
Hy	Hypersthene	SS	Solid solu-
Fs	Ferrosilite		tion
Fo	Forsterite	S	Entropy
Fa	Fayalite	H	Enthalpy
Ol	Olivine	G	Free energy
Ak	Akermanite	E	Activation
Mel	Melilite		energy
Sm	Sodamelilite	A	Activity
Ga (Gr)	Garnet	F	Fugacity
Cor	Corundum	C_P	Heat-capacity

Chapter 1 Introduction

1.1 Basalt and its Relation to Leucite-Bearing Mafic and Ultramafic Rocks

Basaltic magma is considered by many petrologists to be the igneous material derived from the earth's mantle. It thus provides the most direct contact with the nature of the earth at depth and the processes that occur there. For this reason and because basaltic magma is considered to be parental to many other magma types, it has been the subject of numerous studies. Two main basaltic magma types and their contrasting derivatives have been generally recognized: tholeiite basalt (SiO_2-saturated and over-saturated), rhyolite series and alkali basalt (SiO_2-undersaturated), trachyte series. In orogenic belts there is another volcanic rock series, called the calc-alkaline series. However, there are certain potassium-rich mafic and ultramafic rocks, which do not lie in the direct line of descent of the tholeiite, alkaline basalt, or calc-alkaline series. In contrast to basalts these rocks are unusually high in their potassium content. While some mugearites, picrites, monzonites, and alkaline olivine basalts have K_2O contents as high as 1.7% to 2.19% (Table 1.1), the potassium content of most basalts is usually less than 1% (Yoder and Tilley 1962). This is in contrast to the average K_2O content of 90 randomly selected analyses of potassium-rich mafic and ultramafic volcanic rocks, which is 5%; leucite is thus a modal mineral in these rocks. The K_2O contents of the rocks of southern Spain (Borley 1967). Manchuria (China; Ogura et al. 1939), E. Eifel (West Germany; Duda and Schminke 1978), Birunga (east and equatorial Africa; Holmes and Harwood 1937; Higazy 1954; Ferguson and Cundari 1975), and New South Wales (Australia, Cundari 1973) vary between 3% to 6% (Table 1.2); those of the rocks from Leucite Hills (Wyoming, U.S.A.; Yagi and Matsumoto 1966; Carmichael 1967) and West Kimberley (West Australia; Wade and Prider 1940; Prider 1960) are quite high (9% to 11%; Table 1.2) and those of the Italian rocks from such localities as Somma-Vesuvius (Savelli 1967), Roccamonfina (Appleton

Table 1.1. The SiO_2, K_2O and Na_2O contents and SiO_2/K_2O and K_2O/Na_2O ratios of rocks of alkali basalt family

Locality	Rock type	SiO_2	K_2O	SiO_2/K_2O	Na_2O	K_2O/Na_2O	Sources
Nandawar volcano, New South Wales, Australia	Olivine basalt	47.51	1.08	43.99	3.47	0.31	1
Moroto, eastern Uganda	Olivine basalt	45.2	0.81	55.80	3.2	0.25	2
Ard Bheinn, Arran	Olivine basalt	45.52	0.62	73.42	1.86	0.33	3
Fingal's Cave, Staffa	Olivine basalt	47.90	0.53	90.38	2.81	0.19	4
Skye	Mugearite	49.68	1.90	26.14	5.78	0.33	5
Haleakala, Maui, Hawaii	Hawaiite	47.26	1.40	33.76	3.50	0.4	6
Morotu, Sakhalin	Monzonite	53.2	2.19	24.3	6.57	0.33	7
Gough Island	Olivine basalt	47.73	1.70	28.08	2.89	0.59	8
Morotu, Sakhalin	Alkali dolerite	46.56	1.66	28.05	3.67	0.45	9
Clarion Island	Olivine basalt	47.52	1.55	30.66	2.92	0.53	10
Nosappu, Cape, Hokkaido, Japan	Picritic dolerite	46.07	2.01	22.92	2.59	0.78	11
Antrim	Olivine basalt	45.34	0.24	188.9	1.86	0.13	12
East Otago, New Zealand	Alkali olivine basalt	46.29	1.48	31.3	3.9	0.38	13
Tonga	Olivine basalt	50.37	0.31	162.5	1.88	0.17	14
Tahiti	Olivine-augite basalt	45.53	1.05	43.36	2.3	0.46	15

(1) Abbott 1969, (2) Varne 1968, (3) King 1955, (4, 5) Tilley and Muir 1962, (6) Macdonald and Katsura 1964, (7, 9) Yagi 1953, (8) Le Maitre 1962, (10) Bryan 1967, (11) Yagi 1969, (12) Patterson 1951, (13) Coombs and Wilkinson 1969, (14) Bryan et al. 1972, (15) McBirney and Aoki 1968

1972), and Vico (Cundari 1975) range between 6% to 9%. The magnesium contents of the leucite-bearing rocks are variable. While the MgO contents of the rocks from New South Wales, West Kimberley, Leucite Hills, and southern Spain are high and approach the magnesia content of kimberlites, those of rocks from other localities are similar to basalts and these potassic rocks belong to the mafic variety. The SiO_2/K_2O ratios (wt.%) of leucite-bearing rocks are also quite low (usually below 15, Table 1.2) in contrast to those of the rocks of the alkali basalt family (usually > 22, Table 1.1). The other characteristic feature of these rocks is their high K_2O/Na_2O ratio (> 1), which is above 8 or 20 in rocks from Leucite Hills and West Kimberley. In alkali basalts this ratio is less than 1 (Table 1.1). In the normal fractionation scheme of basalts, both potassium and sodium contents of the magmas increase along with their SiO_2 contents and greater enrichment of K_2O over Na_2O does not seem possible by such a process. In highly fractionated rocks, belonging to the basaltic series, plagioclase becomes sodium-rich. However, plagioclases in leucite-bearing rocks are always calcium-rich (Shand 1943). These potassium-rich mafic and ultramafic rocks thus form an independent group of rocks, quite different from the tholeiite and alkali basalt series.

1.2 Mineralogical Composition and Other Geochemical Peculiarities of Leucite-Bearing Rocks

The high alkali and low silica contents of leucite-bearing rocks are reflected by the common occurrence of such minerals as leucite (in some rare instances kalsilite), nepheline, and melilite. Olivines are rich in the forsterite component. Chemistry of the pyroxenes is complex; but usually they have compositions of diopside, salite, and titanaugite. Plagioclase feldspars are calcium-rich and may have an incompatible relationship with melilite (Yoder and Schairer 1969). Although feldspars are absent in rocks of some areas, such as Toro-Ankole and West Kimberley, both plagioclase and K-feldspar are quite common in other areas. Phlogopite and K-rich amphiboles are frequently found in these rocks. Lava flows containing phenocrysts of biotite with minor leucite (glimmerite, Holmes and Harwood 1937) are also known. Besides, mica and amphibole often occur in the xenoliths of potassic lavas in association with pyroxene and olivine.

Other interesting features of these rocks include their high concentration of such elements as Ni, Cr, Ba, Rb, Zr, Nb, La, and Y (Higazy 1954).

Table 1.2. Average SiO_2, K_2O, and Na_2O contents and SiO_2/K_2O and K_2O/Na_2O ratios of leucite-bearing rocks from various localities

Locality	Number of analyses	SiO_2	K_2O	SiO_2/K_2O	Na_2O	K_2O/Na_2O	No.
Central and equatorial Africa	7	43.75 $\rho^a = 4.35$	4.46 $\rho = 1.23$	10.41 $\rho = 2.59$	3.20 $\rho = 1.01$	1.42 $\rho = 0.68$	1
West Kimberley, Australia	7	49.01 $\rho = 6.47$	9.97 $\rho = 1.73$	4.93 $\rho = 0.51$	0.92 $\rho = 0.32$	22.3 $\rho = 7.5$	2
New South Wales, Australia	5	44.05 $\rho = 1.49$	5.48 $\rho = 1.10$	8.24 $\rho = 2.87$	1.26 $\rho = 0.69$	4.65 $\rho = 2.04$	3
Java and Celebes, Indonesia	6	48.83 $\rho = 3.1$	5.51 $\rho = 0.78$	8.79 $\rho = 1.04$	2.43 $\rho = 1.03$	2.12 $\rho = 0.53$	4
Leucite Hills, Wyoming	7	50.45 $\rho = 3.78$	10.82 $\rho = 1.16$	7.68 $\rho = 3.10$	1.50 $\rho = 0.80$	8.15 $\rho = 2.93$	5
Navajo-Hopi, Arizona	6	50.10 $\rho = 3.13$	5.43 $\rho = 1.27$	9.83 $\rho = 2.32$	2.80 $\rho = 0.73$	2.40 $\rho = 0.78$	6
East Eifel, West Germany	7	43.62 $\rho = 0.06$	3.27 $\rho = 0.18$	13.37 $\rho = 0.49$	2.90 $\rho = 0.37$	1.1 $\rho = 0.08$	7
Vico, Italy	7	52.60 $\rho = 2.22$	8.05 $\rho = 1.23$	6.66 $\rho = 0.90$	2.01 $\rho = 0.53$	4.69 $\rho = 1.06$	8
Somma-Vesuvius, Italy	7	51.65 $\rho = 3.47$	6.27 $\rho = 1.04$	8.93 $\rho = 2.69$	3.34 $\rho = 0.68$	1.89 $\rho = 0.44$	9
Roccamonfina, Italy	5	53.34 $\rho = 4.04$	8.59 $\rho = 0.90$	5.36 $\rho = 0.80$	3.11 $\rho = 0.74$	2.98 $\rho = 0.75$	10
Almaria, Spain	7	50.37 $\rho = 3.93$	3.8 $\rho = 0.52$	12.13 $\rho = 2.59$	2.40 $\rho = 0.88$	1.83 $\rho = 0.90$	11
Manchuria, China	6	47.90 $\rho = 3.04$	3.96 $\rho = 1.28$	13.73 $\rho = 5.67$	3.66 $\rho = 0.47$	1.11 $\rho = 0.33$	12

Tristan da Cunha	3	48.37 $\rho = 1.01$	3.55 $\rho = 0.40$	13.65 $\rho = 0.66$	4.60 $\rho = 0.37$	0.76 $\rho = 0.04$	13
Utsuryo Island, Korea	2	57.24 $\rho = 0.67$	6.37 $\rho = 0.31$	9.00 $\rho = 0.35$	5.36 $\rho = 0.44$	1.7 $\rho = 0.37$	14

Sources of data are Tables 5.1, 5.2, 5.3, 5.4, 5.5, 5.6, 5.7, 5.8, 5.9, 5.10, 5.14, 5.15, and 5.16

[a] ρ indicates standard deviation

1.3 A Brief Summary of the Hypotheses of Origin of
Leucite-Bearing Rocks

It is clear from the facts discussed in the first section that leucite-bearing mafic and ultramafic rocks cannot be derived from an alkali basalt-
trachyte series by the process of differentiation and fractionation. Because of this, and because of their predominant occurrence in continental areas, petrologists such as Gorai (1940) and Turner and Verhoogen
(1960) preferred the hypothesis of assimilation of granitic rocks by alkali basalts to explain the origin of some potassium-rich mafic and ultramafic rocks. However, on the basis of Sr isotopic studies of leucite-
bearing rocks Bell and Powell (1969) ruled out such a hypothesis. The
higher concentration of Ni and Cr in these potassic rocks also goes
against such an idea.

Rittmann (1933) invoked the limestone assimilation hypothesis of
Daly (1910) to explain the undersaturated character of these rocks from
the Somma-Vesuvius volcano, which is suggested from the presence of
limestone xenoliths. In these lavas such elements as Ba, Rb, Sr, and Zr
can be supplied from a parent trachyte magma as suggested by Rittmann
(1933), but higher contents of Ni and Cr in these rocks remain unex-
plained, as they cannot be supplied by either a limestone or a trachyte
magma. Besides, in some areas such as Birunga (Holmes and Harwood
1937) and the Hopi province of Arizona (Williams 1936), evidence of
limestone assimilation is lacking. Bowen (1928) and later Wyllie (1974)
considered that such an assimilation hypothesis would require a considerable amount of superheat, but subalkaline magma probably are
not superheated.

Holmes (1950) suggested assimilation of granitic rocks by a carbonatite magma as an alternative hypothesis for the origin of these rocks.
Trace element geochemistry of both granite and carbonatite supports
this hypothesis (Higazy 1954). However, this hypothesis does not explain the genesis of carbonatite magma, which itself poses difficult
petrological problems. It is also questionable that carbonatite magmas
were considerably superheated so as to produce a secondary potassium-
rich magma, which would still retain enough heat to be able to under-
go fractionation and produce different varieties of leucite-bearing
rocks.

The occurrence of leucite-bearing rocks from oceanic environments
has been reported from Utsuryo Island (Tsuboi 1920; Harumoto 1970)
and Tristan da Cunha (Baker et al. 1964), where granitic rocks are

completely lacking. Origin of leucite-bearing rocks independent of granite assimilation is therefore possible.

1.4 Objective of This Study

The main purpose of the present investigation is to examine the evidence for the formation of a potassium-rich magma (capable of producing leucite-bearing rocks near the surface) from the mantle, through an experimental approach. Any such attempt will require the study of multicomponent systems involving all the essential mineral phases present in leucite-bearing rocks. Almost all possible combinations of mineral assemblages of leucitic rocks occur in the following systems: (1) forsterite-diopside-akermanite-leucite, (2) diopside-nepheline-akermanite-leucite, (3) forsterite-diopside-leucite-anorthite, and (4) diopside-leucite-anorthite-SiO_2. Discussions of the phase equilibria study of the systems 1–4 are important to establish the paragenetic relationships of various simplified leucite-bearing rocks. In addition, the systems leucite-albite-anorthite, nepheline-leucite-anorthite, and $(Di_{28}Ne_{29}Lc_{43})_{100-x}-An_x$ are described in the present study in order to understand the cause of the incompatible relationship between the mineral pairs, leucite-albite and melilite-plagioclase. This study also includes the stability of iron leucite ($KFeSi_2O_6$) in leucite and conversion of the latter phase to analcite by sea water. Investigations of various petrologists on the origin of pseudoleucite are also discussed. Bulk compositions of many phlogopite-bearing leucitic rocks lie in the system diopside-forsterite-kalsilite-SiO_2-H_2O. Some joints of the system such as forsterite-kalsilite-SiO_2-H_2O (Luth 1967), phlogopite-enstatite-H_2O, and phlogopite-diopside-H_2O (Modreski and Boettcher 1972, 1973), are therefore discussed. These joins and the results of linvestigations on phlogopite stability (Yoder and Kushiro 1969; Forbes and Flower 1974) are described, as phlogopite is considered to be the most important mineral in connection with the genesis of potassium-rich magmas (Yagi and Matsumoto 1966; Carmichael 1967; Gupta et al. 1976; Beswick 1976). Suggestion has been made that liquid of higher K_2O/Na_2O ratio may be generated by separation of eclogitic material from a picritic magma at depth (O'Hara and Yoder 1967). Experimental study of two picrites at different temperatures and pressures (up to 30 kb P_{H_2O}) is also included to evaluate this hypothesis. Some selective compositions have also been studied at different temperatures and pressures in presence of excess water (up to 25 kb) to determine the nature of the parent materials of leucite-bearing magmas at depth. The results of present experimental studies, as well as those by various other workers, are discussed along with geochemical and isotopic data to elucidate the origin of leucite-bearing mafic and ultramafic volcanic rocks.

Chapter 2 Nomenclature and Petrography of Leucite-Bearing Rocks

2.1 Nomenclature

The nomenclature of the leucite-bearing rocks is rather complex as there was a tendency to assign local names to rocks having slightly different mineralogy. Whenever possible, a simplified nomenclature, based on the mineral components rather than the local names, is used here. Because of the complex old terminology a discussion of old and new nomenclature is necessary. The modal and mineralogical compositions of various leucite-bearing rocks are summarized in Table 2.1. In the following pages they are described in the order of increasing number of main constituent minerals.

2.1.1 Italite. Washington (1927) introduced the term italite for an effusive rock containing 90% or more leucite with minor amounts of augite, glass, and other phases. This term is retained here (Fig. 2.1, Table 2.1).

2.1.2 Leucitite. According to Washington (1927), Zirkel (1867) first proposed the term leucitite for a rock containing leucite and augite in equal amounts with a small amount of nepheline (example Mt. Jugo, Italy; Fig. 2.1, Table 2.1).

Washington proposed the term *albanite* (named after the Albano volcano in Italy) instead of leucitite for an effusive rock composed of equal amounts of leucite and augite. However, the term leucitite of Zirkel is more familiar in the literature and is therefore used here. In modern literature the term *nephelinite* is used in the same sense as leucitite, i.e., for a rock consisting of roughly equal proportions of nepheline and augite.

Weed and Pirsson (1896) suggested the term *missourite* for a group of rocks consisting essentially of augite, olivine, and leucite with minor amounts of biotite, analcite, zeolite, and iron ore (Table 2.1). Washington (1927), however, suggested that olivine is not an essential mineral

Table 2.1. Modal composition of some leucite-bearing rocks (vol %)

Rock name	Locality	Leucite	Melilite	Clinopyroxene	Olivine	Mica	Groundmass	Others[a]
Leucitite[1]	Mt. Jugo, Bolsena, Italy	32	–	50	–	–	–	18
Missourite[2]	Highwood Mountains, Montana	16	–	50	15	6	–	8
Missourite[3] (Type A)	Villa Senni, Italy	49	–	43	–	4	–	4
Missourite[3] (Type B)	Villa Senni, Italy	40	–	35	–	5	–	14
Wyomingite[4]	Leucite Hills, USA	20.3	–	46.1	18.9		14	14.7
Wyomingite[5]	Leucite Hills, USA	40.3	–	39.6	–	17.8	–	2.1
Orendite[5]	Leucite Hills, USA	34.5	–	9.7	–	17.7	–	34[b]
Italite[3]	Villa Senni, Italy	93.5	–	3.0	–	0.5	–	3
Italite[3]	Villa Senni, Italy	90	2	4	–	0.2		3.8
Vesbite[3]	Villa Senni, Italy	60	23	16	–	–		1
Vesbite[3]	Villa Senni, Italy	60	18	20	–	–		2
Cecilite[3]	Villa Senni, Italy	60	23	16	–	–		1
Cecilite[6]	Capo di Bove, Italy	42	12	23	7	–		5
Katungite[6]	Toro-Ankole, Uganda	25	40	–	22	–		13

Katungite[6]	Western flow, Uganda	16	40	–	22.33	–		15.37
Venanzite[6]	Sanvenanzo, Umbria, Italy	30	47	–	17	4		2
Wolgidite[7]	West Kimberley, Australia	27	–	4	–	9	32[c]	28[d]
Fitzroyite[7]	West Kimberley, Australia	47	–	–	–	6	43[c]	4
Cedricite[7]	West Kimberley, Australia	60	–	8-10	5	2	14[c]	9[d]
Mamilite[7]	West Kimberley, Australia	54	–	–	$1\frac{1}{2}$	$3\frac{1}{2}$	18[c]	23[d]
Jumillite[8]	Mauricia, Spain	34	–	14.2	17	9.8	–	25
Wyomingite[8]	Leucite Hills, USA	20.1	–	–	trace	–	79.9	–
Olivine[8] orendite	Leucite Hills, USA	16.1	–	–	3.9	–	80	–
Olivine[8] orendite	Leucite Hills, USA	14.3	–	–	2.3	–	83.4	–
Leucite[9] Nepheline Basanite	Veitskoff, East Eifel; West Germany	e		40	6	3	51	
Leucite[9] Nepheline Tephrite	Eppelsberg, East Eifel, West Germany	e		21	3	1	75	

Sources of data: (1) Washington 1906, (2) Weed and Pirsson 1896, (3) Washington 1927, (4) Weed and Pirsson 1896, (5) Yagi and Matsumoto 1966, (6) Holmes and Harwood 1937, (7) Wade and Prider 1940, (8) Carmichael 1967, (9) Duda 1975

[a] Others refer to the presence of Fe–TiO_2, hauyne, zeolite, apatite, calcite etc. in minor amounts
[b] It also includes 4.1% sanidine
[c] Serpentinite base
[d] Magnophorite and rutile are present
[e] Presence of the mineral in the groundmass

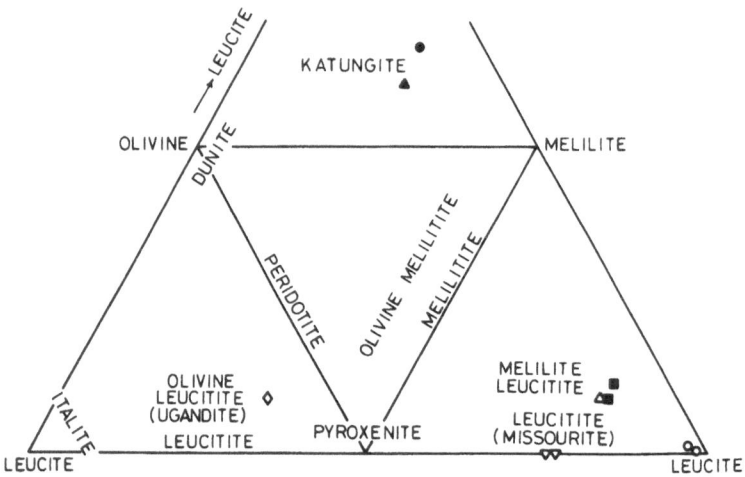

Fig. 2.1. Positions of different types of potassium rich mafic and ultramafic rocks in the tetrahedron forsterite-diopside-akermanite-leucite. ○ Italite; ▽ missourite (Washington 1927); ◇ missourite (Weed and Pirson 1896); △ cecilite; ▲ katungite; ● venanzite; ■ vesbite. (After Gupta 1972)

in missourite and listed the modal percentages of two missourites which are devoid of olivine (Table 2.1). The compositions of these rocks lie in the field of leucitites (Fig. 2.1). In many localities of Uganda and Italy, leucitites are associated with tephrites, basanites, phonolites, and nepheline-bearing rocks; however, in the New South Wales area of Australia, they form an independent group of rocks.

2.1.3 Olivine Leucitite. This term is used in preference to leucite basalt to designate a melanocratic rock, consisting of olivine, clinopyroxene, and leucite (Fig. 2.1). Holmes and Harwood (1937) preferred to use the term ugandite when there was a significant amount of olivine. If the proportion of leucite was greater in an olivine leucitite, it was called *mikenite* by Finckh (1912). Ferguson and Cundari (1975) preferred to call ugandite with a color index in the range of 60 to 75 a mela-basanite.

2.1.4 Melilite Leucitite. Washington (1927) proposed the term *vesbite* for an effusive rock, containing 60% to 70% leucite and about 20% each of augite and melilite (Table 2.1). He also revived the term *cecilite* of Cordier (1901, cited by Washington 1927) to designate a rock also containing leucite, augite, and melilite (Table 2.1) in slightly different proportions. In the triangular diagram of clinopyroxene, melilite, and leucite (Fig. 2.1), it can be seen that the compositions of both vesbite and cecilite fall very close to each other. These rocks are described here as *melilite leucitite*.

2.1.5 Katungite. Holmes (1950) introduced the term *katungite* (after the extinct Katunga volcano, of which it is the sole product for a rock consisting of olivine, melilite, and leucite with minor amounts of apatite, phillipsite, natrolite, fluorite, and rare nepheline (Table 2.1). The name *venanzite* was given to a rock which has the mineralogy of a katungite but is poorer in olivine and richer in leucite (Fig. 2.1, Table 2.1). Only the name katungite is retained here.

2.1.6 Leucite Tephrite. A rock consisting of leucite, augite, plagioclase, and sanidine is called *leucite tephrite.* When the amount of leucite is significant it was called *vesuvite* by Lacroix (cited by Holmes and Harwood 1937). Savelli (1967), Cundari and Mattias (1974), and Ferguson and Cundari (1·975) used the modal components A (alkali feldspar)-P (plagioclase)-F (feldspathoid) to classify feldspar-bearing leucitic rock types into leucititic tephrite, tephritic phonolite, phonolitic tephrite, phonolitic leucitite, tephritic leucitite, etc. (Fig. 2.2). Various combina-

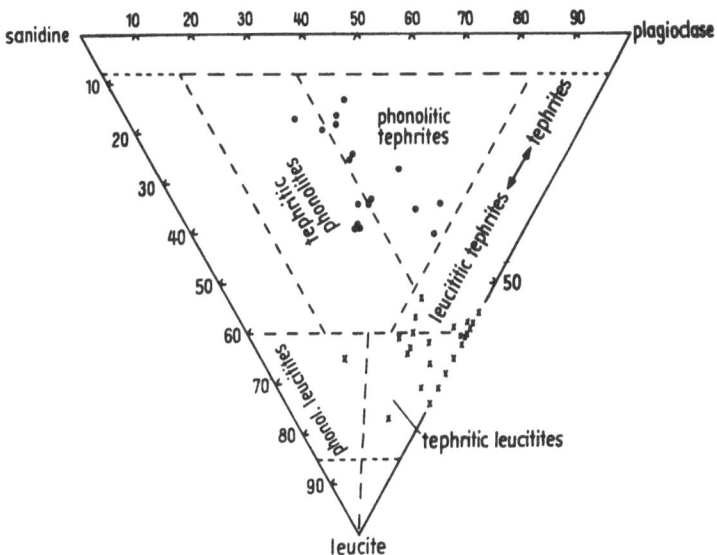

Fig. 2.2. A (alkali feldspar)-P (plagioclase)-F (feldspathoid) diagram for leucite-bearing rocks of the Somma-Vesuvius region. Modal compositions in vol. %. ● Somma lavas; x Vesuvian lavas. (After Savelli 1967)

tions of these rocks have been described by Ferguson and Cundari from the Bufumbira region of Uganda. The rock types shown in Fig. 2.2 have been noted by Savelli (1967) from the Somma-Vesuvius region and by Cundari (1975) from the Vico area, both in Italy. Bulk composition of these rocks would lie in the simplified system diopside-leucite-anorthite-SiO_2 (Chap. 9.2).

2.1.7 Leucite Basanite. The essential minerals of this rock include leucite, olivine, calcic plagioclase, and clinopyroxene. The term *kivite* was used by Holmes and Harwood (1937) to describe a leucite basanite with higher K_2O/Na_2O ratio. The term kivite is not used here. Bulk composition of these rocks would lie in the system forsterite-diopside-leucite-anorthite (Fig. 2.3). Apart from the Bufumbira region, the intimate association of leucite tephrite and leucite basanites is found in the East Eifel region of West Germany (Table 5.11) and different localities of Italy (Chap. 4.9).

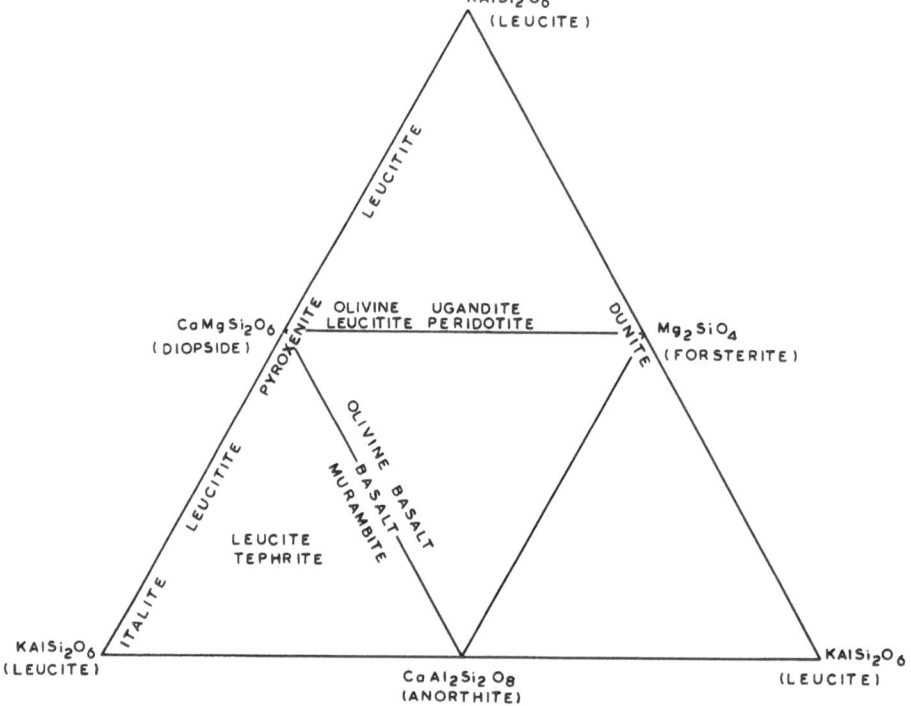

Fig. 2.3. Plot of the composition of leucite-bearing rocks in the system forsterite-diopside-leucite-anorthite

2.1.8 Melilite-Nepheline Leucitite. Holmes and Harwood called a leu-cite-bearing ijolite *niligongite,* which is essentially composed of leucite, nepheline, clinopyroxene, and melilite with accessory olivine. Instead of the term niligongite a more generalized term such as *melilite-nepheline leucitite* is used here. If olivine is present as an essential mineral, it should be called an *olivine-melilite-nepheline leucitite.* Such rocks are found in some localities of the East Eifel region and in the Nyira-gongo area of the Birunga Province of Uganda. The bulk composition of these rocks lies in the system diopside-nepheline-leucitite-akermanite (melilite and olivine appear as reaction products of diopside and nephe-line, as indicated by Schairer et al. 1962).

2.1.9 Jumillite consists of olivine, chromium-bearing diopside, titanif-erous phlogopite, and small amounts of leucite (Borley 1967). Kata-phoritic amphibole is also present as an interstitial mineral.

2.1.10 Verite. This rock contains olivine and phlogopite with small amounts of leucite. Potash feldspar may or may not be present. Calcite is often present as an accessory phase. Type localities of jumillites and *verites* are the Mauricia and Almaria provinces of Spain.

2.1.11 Wolgidite. Various leucite-bearing rocks from the West Kimber-ley Province of Australia were termed by Wade and Prider (1940) *wol-gidite, fitzroyite, mamilite,* and *cedricite,* the definitions of which are given below. *Wolgidite* contains diopsidic pyroxene, large grains of pot-ash richterite, phlogopite, and somewhat turbid crystals of leucite (Ta-ble 2.1). Relic olivine crystals are present in a serpentinous groundmass. Amphibole is more common than phlogopite. Zeolite and ilmenite are present.

2.1.12 Fitzroyite is a fine-grained rock, containing phenocrysts of phlogopite and leucite in a brown glassy groundmass. Rounded to sub-rounded leucite crystals contain characteristic tiny inclusions in the centers of the grains; this mineral is also present in the groundmass. Phlogopite (sometimes twinned) and abundant rutile needles are pres-ent. The amount of leucite far exceeds that of phlogopite (Table 2.1) and quartz amygdules are quite common.

2.1.13 Mamilite is a fine-grained rock, consisting essentially of leucite and potash richterite with rutile needles and very rare phlogopite (Table 2.1). Mica is present only in the groundmass. Leucite crystals are often altered, sometimes containing characteristic inclusions.

2.1.14 Cedricite is a rock rich in leucite with small amounts of diopsidic augite. The brown glassy groundmass contains small to rare amounts of potash richterite, phlogopite, and rutile with serpentines, pseudomorphous after olivine (Table 2.1).

2.1.15 Wyomingite. Cross (1897) coined such terms as *wyomingite, orendite,* and *madupite* for leucite-bearing rocks from the Leucite Hills. Wyomingite is composed of leucite, augite, phlogopite, apatite, calcite, and magnetite in order of abundance (Table 2.1). Phlogopite occurs as phenocrysts, whereas the other minerals are present in the groundmass. Potash richterite is sometimes present as an accessory phase. Wyomingite often shows parallel flow banding, due to the planar arrangements of the phlogopite flakes with vesicular cavities, developed parallel to the flow banding (Yagi and Matsumoto 1966).

2.1.16 Orendite is composed of leucite, sanidine, phlogopite, augite, calcite, and glass. The glass is filled with microlites of sanidine and augite. Sanidine forms euhedral tabular crystals, usually less than 0.1 mm in size, commonly twinned on Carlsbad law. The rocks vary in their crystallinity from a highly glassy variety to a rather crystalline type. In the glassy variety, phlogopite phenocrysts are set in a glassy groundmass, composed of pale brownish glass, filled with abundant acicular crystals of amphibole, less than 0.3 mm in length, which surround the microlite cavities.

2.1.17 Madupite is a rock consisting of small phenocrysts of diopside, enclosed by large poikilitic crystals of phlogopite. The micas are usually set in a turbid groundmass of diopside and chlorite, together with other accessories. Magnetite is present and olivine is absent (Carmichael 1967). Pilot Butte, Leucite Hills, is the type locality for madupite. Bulk compositions of wyomingite, orendite, and madupite can be plotted in the system forsterite-diopside-kalsilite-SiO_2 (Carmichael 1967).

2.1.18 Potassium-Rich Ankaratrite. Sahama (1952) described *potassium-rich ankaratrite* from the west side of Nyamunuka crater in the Katwe-Kikorongo volcanic field of southwestern Uganda. These rocks consist of clinopyroxene, olivine, nepheline, leucite and biotite with apatite, perovskite, analcite, sodalite, and opaque minerals as accessories.

2.1.19 Mafurite. A rock consisting of kalsilite, augite, and olivine was called *mafurite* by Holmes (1950), the type locality of which is Birunga, Uganda.

2.1.20 Shonkinite consists of augite, potash feldspar, biotite, and rare plagioclase. Although it is free from leucite, shonkinite often contains pseudoleucites, e.g., from Shonkin Sag, Montana (Weed and Pirsson 1896; Nash and Wilkinson 1970, 1971) and Gentungen, Indonesia (Iddings and Morley 1915).

2.1.21 Absarokite, Banakite, and Shoshonite. Absarokite is a porphyritic rock consisting of olivine and augite in a groundmass of labradorite with orthoclase rims, olivine, augite, and leucite (Johansen 1931; Vol. 4, pp. 239 and 280). The rock is named after the Absaroka Range, near Yellowstone Park, USA, where it was first described.

Banakite is similar in mineral composition to absarokite, but contains less augite and olivine. It may contain quartz, but in this case olivine is absent.

Shoshonite, named after the Shoshone River in Yellowstone Park is darker in color than the other two rock types described above, but has much the same composition. It may, however, contain some dark glass.

Chapter 3 Mineralogical Composition of Leucite-Bearing Rocks

The most commonly occurring minerals found in these leucite-bearing rocks are leucite, clinopyroxene, plagioclase, potash feldspar, nepheline, and olivine. When plagioclase is present, melilite is absent. Phlogopite and potash richterite are fairly common. Accessory minerals include: biotite, perovskite, wadeite, priderite, apatite, magnetite, ilmenite, and spinel. Different combinations of the above minerals are present in the various types of leucite-bearing rocks. The phase chemistry and chemical variation of these minerals are described below. The phase equilibrium relations among these minerals will be discussed in the later sections.

3.1 Leucite

The composition of natural leucites shows that they usually do not vary much from the ideal formula $KAlSi_2O_6$ (Deer et al. 1963) and their Si/Al ratio is close to 2/1. In natural leucites the main replacement of potassium is found to be sodium. Chirvinsky (1953; cited by Deer et al. 1963) analyzed fourteen leucites from the Vesuvius area and found that the average ratio of K/Na is 86.6/15.4. Tilley (1958) found that leucites from Vogelsberg (FRG) contained 1.12% Na_2O. Fudali (1963) demonstrated that the maximum amount of $NaAlSi_2O_6$ that can be incorporated into the leucite structure is 28 wt. % at 1 kb P_{H_2O} and 800 °C, whereas under similar conditions leucite may contain up to 8 wt. % $KAlSi_3O_8$. After analyzing natural leucites from different localities, he found that their Na_2O content is low, and the maximum amount of $NaAlSi_2O_6$ in leucite was only 10 wt. %. Leucite does not incorporate even 5 wt. % of $NaAlSi_3O_8$ in solid solution at 800 °C and 1 atm (Gupta and Edgar 1975). Schairer (1948) showed that complete series of solid solution exist between the compounds $KAlSi_2O_6$ and $K_2O \cdot MgO \cdot 5 SiO_2$. However, analyses of natural leucites show that the MgO content of leucites is negligible. Analyses of Carmichael (1967;

Table 3.1) show that except for the twinned leucites from West Kimberley, the leucites from the Leucite Hills are slightly nonstoichiometric and they may contain up to 5% potassium disilicate and 2.2% Fe_2O_3. Experimental investigation on the leucite-$KFeSi_2O_6$ system is described later. In Vico also, Cundari (1975) found an excess of silica in leucite (Table 3.1), which he considered to be due to the extended solubility of $KAlSi_3O_8$ and/or differential loss of potassium from the leucite structure. An excess of silica in leucite has also been observed in the Korath range (Brown and Carmichael 1969), New South Wales (Cundari 1973) and Bufumbira (Ferguson and Cundari 1975).

Table 3.1. Analyses of leucites from various localities

	1	2	3	4	5	6
SiO_2	55.4	54.60	54.7	56.2	58.6	54.62
TiO_2	0.2	–	–	–	–	–
Al_2O_3	23.3	22.8	22.4	20.3	18.7	22.93
Fe_2O_3	0.5	2.72	1.0	2.1	2.2	0.26
FeO			–	–	–	0.26
MnO	–	–	–	–	–	
MgO	–	0.03	–	–	–	–
CaO	1.1	0.16	0.01	0.03	0.04	0.08
Na_2O	1.8	1.18	0.01	0.02	0.06	0.66
K_2O	18.3	18.46	21.5	21.1	20.7	21.02
Total	100.6	99.95	99.7	99.8	100.3	99.95

1. Leucite from a leucite theralite, Nyamlagira volcano, Zaire (Bowen and Ellestad 1937). It includes 0.2% H_2O^+
2. Leucite from Roccamonfina, Italy. Anal. B. Kronburg (Gupta and Fyfe 1975)
3. Leucite from a cedricite, West Kimberley, Australia (Carmichael 1967). BaO below 0.02%
4. Leucite from a wyomingite, Leucite Hills, Wyoming, USA (Carmichael 1967). BaO below 0.02%
5. Leucite from an orendite, Leucite Hills, Wyoming, USA (Carmichael 1967)
6. Leucite from a leucitite; Lake Kivu, Zaire (Sahama 1952). It includes 0.12% H_2O^+

In the Leucite Hills, Carmichael (1967) and Sobolev et al. (1975) found that leucite contains variable amounts of fluid inclusions. Similar fluid inclusions within leucite crystals from New South Wales have also been reported by Cundari (1973). Leucite is often characterized by solid inclusions. Tsuboi (1920) noted inclusions of augite microlites and magnetite within leucite crystals from Utsuryo Island. Inclusions of small pyroxene crystals within leucite are found in the rocks from Eifel and other localities.

Leucites very often show slight anisotropism. Those from Australia are often twinned and are slightly anisotropic, in contrast to the completely isotropic ones from the Leucite Hills, which were considered by Carmichael to have been formed above their cubic tetragonal inversion temperature of 630 °C (Faust 1963). Cundari (1973) considered that the large leucite crystals of New South Wales with multiple twinning have been quenched below their inversion temperature. The leucites from Utsuryo Island also show weak birefringence (Tsuboi 1920). All synthetic leucites crystallized in the present study showed weak birefringence (~0.001). Analyses of leucites from various localities are given in Table 3.1.

3.2 Nepheline

Natural nepheline always contains kalsilite in solid solution. Smith and Tuttle (1957) demonstrated that nepheline can contain up to 62.5 wt. % of kalsilite in solid solution but at temperatures below 400 °C the solubility of kalsilite in nepheline decreases to 26 wt. %. In the central African lavas, where nepheline and leucite coexist, Sahama (1952) found that the K_2O content varies from 0.60% to 0.98%. Greig and Barth (1938) showed that nepheline can contain up to 33 wt. % of albite (the composition of such a nepheline would be equivalent to $nepheline_{85}$ $quartz_{15}$) in solid solution. Cundari (1973) found that the nephelines from the leucite-bearing lavas of New South Wales contain up to 8 wt. % of excess silica. All the analyses of nepheline collected by Deer et al. (1963) showed that it contains excess silica. In the system nepheline-anorthite, nepheline incorporates up to 35 mol % $CaAl_2Si_2O_8$ in solid solution (Bowen 1912). Gupta and Edgar (1974) showed that in the system nepheline-leucite-anorthite, both calcium and potassium are equally favored by its structure. However, the calcium content of natural nepheline from leucite-bearing mafic and ultramafic lavas is low. Onuma et al. (1972) showed that nepheline can also incorporate the iron nepheline compound $NaFe^{3+}SiO_4$ up to 25 mol % at 700 °C and atmospheric pressure. Small amounts of Mg, Mn, and Ti are also reported in nephelines (Deer et al. 1963, p. 240). The chemical compositions of some natural nephelines are given in Table 3.2.

Table 3.2. Analyses of nephelines from potassic rocks

	1	2	3	4	5	6
SiO_2	42.28	40.20	40.74	43.55	41.52	43.97
TiO_2	0.07	0.05	0.11	–	–	–
Al_2O_3	33.71	32.51	33.39	34.66	34.09	32.89
Fe_2O_3	0.80	1.82	0.83	–	0.79	–
FeO		0.57				
MgO	0.03	0.10	0.25	0.05	–	–
CaO	0.56	1.44	0.91	4.44	0.06	0.43
Na_2O	16.61	10.86	12.53	12.09	15.76	15.73
K_2O	5.75	12.22	11.13	4.87	6.85	5.45
H_2O^+	0.34	–	0.23	0.25	–	–
H_2O^-	0.03	–	0.06	0.25	0.55	–
Total	100.18	99.77	100.18	100.16	99.65	98.47

1. Nepheline from a nepheline-leucite diabase, Meiches, Vogelsberg, Hessen (Tilley 1958)
2. Nepheline from a potash ankaratrite, Mt. Nyiragongo, Zaire (Sahama 1952)
3. Nepheline from a leucite nephelinite, Lake Kivu, Zaire. Anal. R. B. Ellestad (Bowen and Ellestad 1937)
4. Nepheline from a block containing biotite, hornblende, sanidine, and vesuvianite, Monte Somma, Italy (Banister and Hey 1931)
5. Nepheline from a pegmatite, Bearpaw Mountains, Montana, USA. (Pecora 1942). It includes 0.03% BaO
6. Average of 16 nephelines from Vesuvius, Italy (Chirvinsky 1952; cited by Deer et al. 1963)

3.3 Kalsilite

Tuttle and Smith (1958) found that kalsilite can incorporate slightly above 20 wt. % $NaAlSiO_4$ at 1000 °C. However, with a lowering of the temperature the kalsilite solid solution would exsolve nepheline and at 400 °C and 1 atm the solubility of $NaAlSiO_4$ is less than 5 wt. %. In potassium-rich lavas of Nyiragongo (Zaire, Africa) Sahama (1960) found a perthitic intergrowth of nepheline and kalsilite. Recently similar intergrowth has also been found in leucite-bearing pyroxenite from East Greenland by Gittins (personal communication 1977). Chemical analyses of some natural kalsilites are given in Table 3.3, which shows that the Na_2O contents are considerably less than the K_2O contents of nepheline and that the maximum amount of Na_2O does not exceed 2.09%. However, the presence of Fe_2O_3 in natural kalsilite indicates that the $Al^{3+} \rightleftharpoons Fe^{3+}$ substitution can be up to 5.07% (Deer et al. 1963, p. 252). As the compound $K_2O \cdot MgO \cdot 3 SiO_2$ (Roedder 1951)

Table 3.3. Analyses of kalsilites and kaliophilites from potassic rocks

	1	2	3	4	5	6
SiO_2	38.47	37.98	38.50	39.04	38.0	39.2
TiO_2	–	0.05	0.09			
Al_2O_3	30.81	31.73	26.27	31.96	28.8	33.36
Fe_2O_3	1.63	0.98	5.07	0.98	–	–
FeO	0.26	–	0.53	–	–	–
MnO	–	0.01	0.02			
MgO	0.63	–	0.87	0.15	–	–
CaO	0.20	–	0.44	0.33	0.5	0.47
Na_2O	2.09	0.87	2.07	3.89	trace	2.88
K_2O	25.65	27.99	24.85	22.84	32.2	24.13
H_2O^+	0.20	0.40	1.04	0.60	–	–
H_2O^-	–	–	0.05			
Total	99.94	100.01	99.80	99.79	99.5	100.04

1. Kalsilite from a venanzite, San Venanzo, Italy. Anal. H.B. Wiik (Bannister et al. 1953)
2. Kalsilite from kalsilite fraction of a kalsilite perthite, which is the constituent mineral of a fallen rock near Baruta crater, Nyiragongo, Zaire. Anal. P. Ojaperä (Sahama et al. 1956)
3. Kalsilite from a mafurite, Kyambogo crater, Bunyaruguru, southwestern Uganda (Sahama 1954)
4. Kaliophilite from an ejected block, Monte Somma, Vesuvius, Italy. Anal. N. Sahlbom (Mügge 1927)
5, 6. Kaliophilite from a biotite and augite-bearing block, Monte Somma, Vesuvius, Italy (Bannister and Hey 1931)

is isostructural with kalsilite, the substitution Mg^{2+} Si^{4+} $\rightleftharpoons$ $2\,Al^{3+}$ in kalsilite is possible.

Compared to leucite and nepheline the occurrence of kalsilite is rare. It occurs in association with these minerals in the lavas of Mt. Nyiragongo, Zaire (Sahama 1953b, 1957, 1960), Bunyaruguru (Holmes 1942; Bannister and Hey 1942), Katunga, Uganda (Combe and Holmes 1945) and San Venanzo, Italy (Bannister et al. 1953).

3.4 Kaliophilite

Kaliophilite is a rare mineral, first found in the ejected blocks of biotite pyroxenites or metamorphosed calcareous rocks from Monte Somma, Vesuvius, Italy (Mügge 1927; Bannister and Hey 1931). Tilley and Henry (1953) reported the occurrence of kaliophilite in the ejected blocks of

Albano, Latium, Italy, where it is associated with hedenbergitic pyroxene, grossularite-andradite, leucite, hauyne, and latiumite.

As shown in Table 3.3, kaliophilite has a composition similar to that of kalsilite. Replacement of K by Na and some cation deficiency are commonly observed in the crystal structure of kaliophilite. The maximum ratio of Na/K in this mineral is found to be 1:10. Kaliophilite is hexagonal with $\epsilon = 1.527$ and $\omega = 1.531$. It is regarded as polymorphous with kalsilite, but its structure has not been investigated in detail.

Kaliophilite was synthesized first by Bowen (1928) at dry condition and later by Tuttle and Smith (1958) in the system nepheline-kalsilite from $Ne_{10}Ks_{90}$ composition at temperatures between $1000°$ and $1100\ °C$ under a pressure of 0.5 kb P_{H_2O}. The synthetic kaliophilite differs from the natural ones in having a more disordered crystal structure. According to Tuttle and Smith (1958), kaliophilite is metastable at all temperatures under atmospheric pressure in contrast to kalsilite.

3.5 Sanidine

Although sanidine is absent in many rocks from Toro-Ankole (Africa) and West Kimberley (Australia), it occurs in leucite-bearing phonolites, tephrites, trachytes, and shoshonites. In the Leucite Hills areas it is not found in wyomingites and madupites, but is present in orendites.

Carmichael (1967) studied sanidines from leucite-bearing rocks of Spain, West Kimberley, and the Leucite Hills. In all these sanidines, he found high concentration of the $KFe^{3+}Si_3O_8$ molecule, up to 9.6 wt.%. In the orendites of the Leucite Hills, the content of this molecule in sanidine varies from 10.3 to as high as 18.4 wt.%. The albite content of sanidines in the jumillite rocks of Spain is about 8.8%, whereas it is only 1.2% to 2.7% in the K-feldspars from the Leucite Hills. The CaO content of these feldspars is low, but BaO content is always high. Anorthoclase is associated with the leucite-bearing rocks of Utsuryo island (Harumoto, 1970). Analyses of some K-feldspars from leucite-bearing rocks are listed in Table 3.4.

3.6 Plagioclase

Plagioclase is an important mineral in leucite-bearing tephrites and basanites from Eifel, Birunga volcanic province, Somma-Vesuvius, and

Table 3.4. Analyses of K-feldspars from leucite-bearing rocks of various localities

	1	2	3	4	5	6
SiO_2	63.62	63.9	64.1	64.3	65.0	65.0
TiO_2	0.08	–	–	–	0.12	0.06
Al_2O_3	19.12	16.7	15.6	16.6	19.5	19.8
Fe_2O_3	0.47	2.5	3.0	2.7	0.52	0.23
FeO	–	–	–	–	–	–
CaO	0.05	0.01	0.07	0.06	0.79	0.78
Na_2O	2.66	1.04	0.19	0.19	3.3	2.3
K_2O	12.09	14.09	16.2	16.3	11.0	12.0
BaO	1.56	0.96	0.60	0.79	–	–
SrO	–	0.06	–	–	–	–
Total	99.81	99.26	99.76	100.94	100.23	100.17

1. Sanidine from a leucite-nepheline dolerite; Meiches Vogelsberg, Hessen (Tilley 1954). It also includes 0.11% H_2O and 0.05% MgO. It corresponds to Or_{73}-$Ab_{22.9}An_{0.3}Cn_{3.8}$
2. Sanidine from a jumillite, Spain (Carmichael 1967). It corresponds to $Or_{79.2}$-$(Fe-Or)_{9.6}Ab_{8.8}Cn_{2.2}Sf_{0.2}$
3. Sanidine from an orendite, North Table Butte, Leucite Hills (Carmichael 1967). It corresponds to $Or_{85.1}(Fe-Or)_{11.5}Ab_{1.6}Cn_{1.4}An_{0.4}$
4. Sanidine from North Table Mountain, Leucite Hills (Carmichael 1967). It corresponds to $Or_{86}(Fe-Or)_{10.3}Ab_{1.6}Cn_{1.8}An_{0.3}$
5, 6. Alkali feldspars from leucite-bearing rocks of Vico, Italy. Anal. A.F. Ferguson (Cundari 1975). Or = orthoclase; Fe-Or = iron orthoclase; Ab = $NaAlSi_3O_8$; Cn = $BaAlSi_3O_8$; An = anorthite, Sf = $SrAl_2Si_2O_8$

Vico lavas. Shand (1943) pointed out that plagioclases occurring in leucite-bearing lavas are always rich in anorthite. Later work by Savelli (1967) and Cundari (1975) also supported this conclusion. The petrochemical implications of leucite-albite incompatibility are discussed in a later section.

Savelli (1967) found that the anorthite content of plagioclase in the Somma lavas varied between 68% and 78%, whereas in the case of Vesuvian lavas the anorthite component was higher (82% to 84% by weight).

Plagioclases from the Vico area were studied by Cundari (1975), where they are sometimes included within leucite, although they are also present in notable amounts as groundmass constituents. The plagioclase phenocrysts in this area show concentric, normal, and reversed zoning, and are characterized by high anorthite component (76% to 88.9% by weight). Analyses of plagioclases from this area are given in Table 3.5.

Table 3.5. Analyses of plagioclases from leucite-bearing lavas of the Vico area, Italy

	1	2	3	4	5	6
SiO_2	45.8	46.3	47.4	46.1	47.2	48.1
TiO_2	–	–	0.02	–	–	–
Al_2O_3	34.0	34.0	32.0	33.0	32.0	32.0
Fe_2O_3	0.74	0.75	0.58	0.75	0.73	0.71
MnO	0.06	–	0.04	–	0.02	–
MgO	–	–	0.04	–	0.04	0.04
CaO	18.4	17.5	17.1	18.3	17.8	16.0
Na_2O	1.23	1.50	2.3	1.63	2.3	2.9
K_2O	0.11	0.21	0.35	0.14	0.38	0.15
Total	100.34	100.26	99.83	99.92	100.47	99.90

All six plagioclases are from leucite-bearing lavas of the Vico area, Italy. Anal. A.K. Ferguson (Cundari 1975)

3.7 Pyroxene

Pyroxenes from leucite-bearing rocks are solid solutions of $CaMgSi_2O_6$, $CaFe^{2+}Si_2O_6$, $CaFe_2^{3+}SiO_6$, $CaAl_2SiO_6$, $MgAl_2SiO_6$, $CaTiAl_2O_6$, $NaAlSi_2O_6$, $NaCrSi_2O_6$, $NaFeSi_2O_6$, $Ca_2Si_2O_6$, $Mg_2Si_2O_6$, and $Fe_2Si_2O_6$. Diopside ($CaMgSi_2O_6$), however, is the principal component of the clinopyroxenes and the others are present only in small amounts.

Carmichael (1967) studied the compositions of pyroxenes from Spanish lavas (Jumillite) and volcanic rocks of the Leucite Hills and West Kimberley. Some of his pyroxene analyses are given in Table 3.6. Carmichael found that zoning is absent in these pyroxenes, although they are often slightly iron-rich toward their margins. However, in the case of some of the pyroxenes in madupite he noted minor zoning, with the development of acmitic margins. Carmichael's data show that when the compositions of the pyroxenes are plotted in the diopside-hedenbergite-ferrosilite-enstatite quadrilateral they fall close to the diopside corner.

Clinopyroxenes from the Alban Hills of Italy were studied by Washington and Merwin (1923). These pyroxenes contain 1.19% TiO_2 and their Al_2O_3 and CaO contents are also high. The indices of refraction of these pyroxenes were found by them to be $N_z = 1.727$, $N_y = 1.710$, and $N_x = 1.703$. Yagi and Onuma (1967) studied the system diopside-$CaTiAl_2O_6$ and established that diopside containing 11 wt. % $CaTiAl_2O_6$ has $N_z = 1.718$, $N_y = 1.697$ and $N_x = 1.684$. Gupta et al.

Table 3.6. Analyses of clinopyroxenes from leucite-bearing rocks of various localities

	1	2	3	4	5	6	7	8	9	10	11	12	13	14
SiO_2	54.1	54.5	55.1	52.26	52.7	51.8	52.3	46.4	44.77	39.14	48.91	47.27	47.12	47.75
TiO_2	0.63	0.65	0.45	0.73	1.4	1.5	1.4	0.76	2.95	4.81	1.27	1.52	1.46	1.43
Al_2O_3	0.23	0.76	0.20	1.94	1.2	2.4	1.1	7.1	8.18	10.86	5.95	7.40	7.63	6.43
Fe_2O_3	–	–	–	1.08	–	–	–	–	–	–	2.74	3.07	3.01	2.85
FeO	3.0	2.1	2.9	2.29	3.8	4.4	3.9	10.8	5.93	7.84	4.01	4.91	5.02	5.78
MnO	0.08	0.09	0.09	0.07	0.11	0.05	0.08	0.27	trace	0.15	0.14	0.15	0.14	0.12
MgO	17.1	17.8	17.2	16.73	16.1	15.0	16.2	11.2	12.21	10.26	13.69	12.26	12.83	13.63
NiO	0.03	–	0.02	–	–	–	–	–	–	–	–	–	–	–
CaO	24.2	24.7	24.2	22.80	23.6	23.9	24.4	23.2	23.76	22.96	23.18	22.88	22.82	21.49
Na_2O	0.16	0.24	0.15	0.87	0.30	0.26	0.23	0.36	0.36	0.38	0.41	0.47	0.46	0.49
K_2O	–	–	–	0.18	–	–	–	–	–	–	0.08	0.06	0.03	0.03
Cr_2O_3	0.02	–	0.04	0.54	0.51	0.79	0.52	–	–	0.14	–	–	–	–
Total	99.55	100.84	100.35	99.49	99.72	100.10	100.13	100.09	98.16	96.54	100.38	99.99	100.52	100.00

1. Diopside from an orendite flow; Orenda Butte, Leucite Hills (Carmichael 1967)
2. Diopside from a madupite, west side Pilot Butte, Rock Springs, Leucite Hills (Carmichael 1967)
3. Diopside from a wyomingite, Boars Tusk, Leucite Hills (Carmichael 1967)
4. Clinopyroxene from a jumillite, Spain (Borley 1967)
5–7. Clinopyroxenes from two leucitites, Begargo Hill, New South Wales (Cundari 1973)
8. A pyroxene from a leucite-bearing tephrite from Vico, Italy (Cundari 1975)
9. A clinopyroxene from a leucite nephelinite from Herchenberg, East Eifel, West Germany (Duda 1975)
10. A clinopyroxene from a leucite-nepheline basanite, Nickenicher Sattel, East Eifel, West Germany (Duda 1975)
11–14. Clinopyroxenes from the Somma-Vesuvius region of Italy (Rahman 1975)

(1973a) noted that in the system diopside (Di)-$CaTiAl_2O_6$ (Tp)-SiO_2 (Si), diopside crystallizing from a melt of composition $Di_{60}Tp_{35}Si_5$ at 1200 °C has $N_z = 1.730$ and $N_x = 1.705$. Barth (1933) showed that diopside containing 10 wt. % $CaMgTi_2O_6$ has $N_z = 1.730$, $N_y = 1.708$, and $N_x = 1.695$.

A study of pyroxenes from the leucite-bearing tephrites and basanites of the Vesuvius region by Rahman (1975) showed that the most important characteristic of these clinopyroxenes is the extensive zoning of the phenocrysts and microphenocrysts, sometimes with up to sixty zones in one grain. The amount of silica in these pyroxenes is fairly constant and they are characterized by high Al_2O_3. Thompson (1972) found that the Al_2O_3 variation in some Somma-Vesuvius clinopyroxenes could range between 1 and 9.6 wt. % within a single phenocryst. According to Rahman (1975) the iron content of the Vesuvius clinopyroxenes is also quite high, suggesting that they are rich in the $CaFe_2^{3+}SiO_6$ (Huckenholz et al. 1969) and $CaFe^{3+}AlSiO_6$ molecule (Hijikata and Onuma 1969). The TiO_2 contents of these pyroxenes are relatively low and they plot in the salite field of Poldervaart and Hess (1951). The crystallization trend of the pyroxenes is nearly parallel to the diopside-hedenbergite join and is restricted to the compositional field of ferrosilite$_9$-ferrosilite$_{15}$, wollastonite$_{45}$-wollastonite$_{50}$.

Clinopyroxenes of the leucitic rocks from New South Wales have a very narrow compositional range within the diopside field (Cundari 1975; p. 475). Late crystallizing pyroxenes approach the salite boundary with a concomitant increase in calcium, iron, and sodium content. The TiO_2 contents of the pyroxenes are high (Table 3.6) indicating the probable presence of a $CaTiAl_2O_6$ molecule. Late stage pyroxenes are characterized by enrichment in sodium, titanium, and iron. Ewart et al. (1976) noted that soda-enrichment of the pyroxenes in rocks of the alkalic basalt-rhyolite series from S.E. Queensland and N.E. New South Wales occurs only after iron-enrichment in pyroxenes, and was considered by them to reflect interplay of f_{O_2} and $a_{Al_2O_3}^{liquid}$.

In the clinopyroxenes from the leucite-bearing lavas of the Vico area (Roman Province), Cundari (1975) found that when the compositions of these pyroxenes were plotted in the conventional Ca-Mg-Fe (total) + Mn diagram, they fall in the salite field and show a tendency toward iron enrichment, parallel to the diopside-hedenbergite join with no significant distinction between core and rim compositions. The distribution of titanium with respect to aluminum (z) in coexisting core and rim compositions showed contrasting trends. According to

Cundari this may be related to a complex history of equilibration. He further noted the general absence of normative jadeite and the presence of a significant amount of Ferri-Tschermak's molecule ($CaFe_2^{3+}SiO_6$), ($CaFe^{3+}AlSiO_6$), moderate $CaAl_2SiO_6$ (5.7 to 19.9 wt.%) and an accessory amount of acmite (1.2 to 3.8 wt.%). The $CaTiAl_2O_6$ contents of these pyroxenes are less than 2.5 wt.%. Duda and Schminke (1978) studied the clinopyroxenes from the Laacher See area (Fig. 3.1). The chemical compositions of clinopyroxenes from various localities are given in Table 3.6.

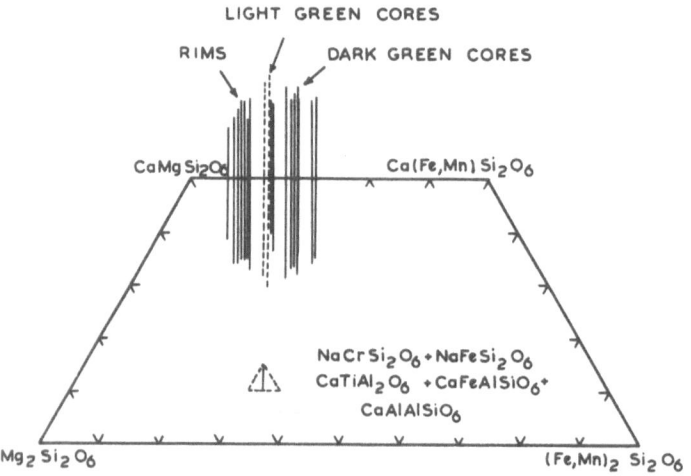

Fig. 3.1. Plot of the composition of clinopyroxenes in the system wollastonite-enstatite-ferrosilite. (After Duda and Schminke 1978)

3.8 Olivine

Natural olivine is mainly a solid solution of forsterite and fayalite (Bowen and Schairer 1935), but may also contain small amounts of monticellite (Ricker and Osborn 1954), larnite (Ca_2SiO_4), tephrite (Mn_2SiO_4), and nickel olivine (Ni_2SiO_4). Borley (1967) determined the forsterite content of olivine from verite and jumillite lavas. She found that in verites the forsterite content varies from 89% to 91.5%, whereas in jumillites it ranges between 85.5% to 89%. Borley (1967) also analyzed olivines in orendite and wyomingite from South Table Mountain and North Table Mountain, Leucite Hills. Her analyses of olivines (Table 3.7) indicate that the forsterite content of olivines varies

Table 3.7. Analyses of olivines from potassic rocks of various localities

	1	2	3	4	5	6	7	8	9
SiO_2	39.42	39.40	39.38	39.0	38.3	40.0	39.3	39.5	39.7
TiO_2	–	–	–	–	–	–	0.03	0.03	0.03
FeO	13.48	13.19	13.86	18.7	19.2	11.6	18.6	12.50	12.9
MnO	–	–	–	0.43	0.27	0.15	0.38	–	0.21
MgO	45.06	45.02	44.75	41.1	41.0	47.2	40.9	47.4	46.9
NiO	0.18	0.18	0.18	0.15	0.20	0.19	0.10	0.28	0.29
CaO	0.29	0.31	0.24	0.29	0.35	0.24	0.42	0.26	0.13
Cr_2O_3	–	–	–	–	0.04	0.05	–	–	0.04
Total	98.43	98.10	98.41	99.67	99.36	99.43	99.76	99.97	100.20

1–3. Olivines from leucite basanites; Rothenberg, East Eifel, West Germany (Duda 1975)

4, 5. Olivines from leucitites; Begargo Hill, New South Wales, Australia (Cundari 1973)

6. Olivine from an olivine leucitite; Condoblin, New South Wales, Australia (Cundari 1973)

7, 8. Olivine from leucitites, Begargo Hill, New South Wales, Australia (Cundari 1973). 7 includes 0.03% Al_2O_3

9. Olivine from a leucitite, Lake Cargelligo, New South Wales, Australia (Cundari 1973)

from 85% to 90%. Carmichael (1967; p. 36) found two modes of occurrence of olivine phenocrysts surrounded by a reaction rim of phlogopite, and small red-rimmed phenocrysts with no evidence of instability. Olivines from jumillite occur as large crystals, surrounded by sanidine and phlogopite. In some cases the olivines are altered to bowlingite. Simkin and Smith (1970) found that the olivines from plutonic rocks contain less than 0.14 wt. % CaO, whereas those from volcanic rocks always contain greater amounts of CaO. White (1966) supported the conclusions of Simkin and Smith from his study of olivines from ultramafic inclusions in Hawaiian lavas. Warner and Luth (1973) demonstrated that the CaO content of olivines systematically decreases at higher pressures. Gupta and Yagi (1978) established that in the system forsterite-grossularite, forsterite contains 5 wt. % of monticellite at 8 kb and 1000 °C, whereas at 20 kb and 1000 °C it contains only 1.56 wt. % of monticellite in solid solution. The CaO contents of olivines studied by Carmichael (1967) show that these different rock types were produced under both plutonic and volcanic conditions.

Cundari (1973) studied the olivines of the olivine leucitites from New South Wales and found that the compositions of olivines vary between Fo_{79-93}, mostly within the range Fo_{82-88} and varying degree of alteration to iddingsite is very common. Except for one example, the CaO contents of the olivines are higher than 0.14 wt. % (Table 3.7).

The compositions of olivines from the Laacher See area (Duda and Schminke 1978) indicate that the CaO content of some of the olivines is also higher than 0.14 wt. %, which was found by Carmichael (1967) to be the lower CaO limit for the olivines from the Leucite Hills. Iddingsitization of olivine is also very common in the Laacher See area.

3.9 Mica

Phlogopite is a very common mineral in the potassic rocks of West Kimberley, Spain, and the Leucite Hills and occurs as accessory mineral in the rocks from other areas. Phlogopite is also commonly found in the xenoliths in association with pyroxene and olivine.

Wade and Prider (1940) studied the phlogopites from West Kimberley. They noted that this mineral occurs as a deep reddish brown variety in the form of pseudohexagonal plates in wolgidites. It invariably has a poikilitic relationship to leucite and clinopyroxene. The pleochroic scheme of the phlogopites is usually, $X \leqslant Y < Z$, but some phlogopites of wolgidites have the scheme, $Y < X < Z$. In the Spanish lavas phlogopite occurs as large crystals, enclosing leucite or its pseudomorphs, apatite, and diopside. Carmichael (1967) noted that the Australian phlogopites are typically zoned whereas in the Leucite Hills, this mineral was found to form a reaction rim around olivine and chromite. Analyses of phlogopites show that they are rich in fluorine: 2.46% (phlogopite from Wyoming), 2.16% (phlogopite from Spain, Osann 1906), and 0.66% (Wyoming, Carmichael). Yagi and Matsumoto (1966) also studied the phlogopites from the Leucite Hills area. They determined the composition of phlogopite from wyomingite as, $(K, Na, Ba, Sr)_{2.00}(Mg, Fe^{2+}, Fe^{3+}, V)_{5.65}(Al, Ti, Cr)_{2.42}O_{21.68}Si_6$ (OH, F, Cl)$_{2.32}$.

The cell parameters of this phlogopite are as follows: $a_0 = 5.33$, $b_0 = 9.32$, $c_0 = 10.25$, $\beta = 99° 56'$. From these data they concluded that the phlogopite belongs to the 1 M polymorph of phlogopite as defined by Smith and Yoder (1956). Table 3.8 shows that the titanium content is high in the phlogopites from West Kimberley, New South Wales, and Spain in contrast to that of the average phlogopites from the Leucite Hills and East Eifel.

Table 3.8. Analyses of phlogopites from leucite-bearing rocks

	1	2	3	4	5	6	7	8	9	10
SiO_2	40.78	41.33	42.8	42.3	38.99	35.63	35.44	35.2	35.0	37.4
TiO_2	8.97	1.45	5.9	3.5	8.43	5.33	5.43	10.4	10.9	9.9
Al_2O_3	10.95	12.91	9.9	8.9	8.20	15.73	16.07	12.8	13.6	11.0
Fe_2O_3	2.18	2.27	–	–	1.24	–	–	–	–	–
FeO	3.73	1.33	6.6	5.5	6.41	8.23	8.38	10.8	12.0	8.2
MnO	trace	–	0.03	0.04	0.03	trace	trace	0.12	0.10	0.05
MgO	19.66	23.94	18.3	22.4	19.31	18.23	17.77	17.0	16.20	19.5
NiO	–	–	0.15	0.13	–	–	–	–	0.04	0.08
CaO	0.11	trace	0.04	0.02	0.60	0.07	0.07	0.03	0.03	–
Na_2O	0.11	0.47	0.44	0.30	0.85	0.65	0.59	0.87	0.76	0.71
K_2O	10.59	9.64	9.7	9.8	8.75	9.10	8.82	7.1	7.2	8.5
H_2O^+	1.87	1.79	–	–	1.34	–	–	–	–	–
H_2O^-	0.19	1.07	–	–	1.85	–	–	–	–	–
BaO	0.35	0.84	1.6	1.6	–	–	–	–	–	–
Cr_2O_3	trace	0.46	0.80	0.07	–	0.11	0.11	–	–	–
F	0.66	1.17	–	–	–	–	–	–	–	–
Total	100.15	99.30	96.26	94.56	96.00	93.08	92.68	94.32	95.83	95.34

1. Phlogopite from a leucite lamproite, West Kimberley. Anal. H.C. Vincent (Wade and Prider 1940)
2. Phlogopite from a wyomingite, Emmons Cone, Leucite Hills (Yagi and Matsumoto 1966). It also includes 0.41% SrO, 0.21% Cl, and 0.008% SO_3
3. Phlogopite from a jumillite, Spain (Carmichael 1967)
4. Phlogopite from a madupite, west side Pilot Butte, Rock Springs, Leucite Hills (Carmichael 1967)
5. Phlogopite from a jumillite lava, Spain (Borley 1967). It also includes 0.15% P_2O_5
6. Phlogopite from a leucite tephrite, Krufter Ofen, East Eifel, West Germany (Duda 1975)
7. Phlogopite from a leucite tephrite, East Eifel, West Germany (Duda 1975)
8, 9. Phlogopites from two leucitites, Begargo Hills, New South Wales (Cundari 1973)
10. Phlogopite from an olivine leucitite, Condoblin, New South Wales (Cundari 1973)

Biotite has been reported by Holmes and Harwood (1937) and Cundari (1973) in leucite-bearing lavas from Birunga and Roman Province, respectively. In these areas it occurs in xenoliths and sometimes in lavas, consisting entirely of biotite phenocrysts with minor leucite (biotitite or glimmerite). Holmes (1950) described xenoliths consisting of olivine, biotite, and pyroxene (O.B.P. series) associated with katungite.

3.10 Potash Richterite

Potash richterite (magnophorite) has been reported from the jumillite lavas (Spain), leucite-bearing rocks of New South Wales, West Kimberley, and the Leucite Hills. Table 3.9 shows that all potash richterites are rich in Ti and Mg. Sr and Ba contents are also high. Wade and Prider (1940) studied the potash richterites from West Kimberley, where they occur as skeletal or occasionally as idiomorphic plates. The pleochroic scheme is as follows: Y (reddish, like hypersthene) > Z (yellow) > X (pale yellow or colorless).

In New South Wales the amphibole (β = 1.641–1.651; γ = 1.646–1.652) occurs as an accessory groundmass mineral (Cundari 1973). They often contain needle-like inclusions, arranged in parallel patterns. Cundari considered these amphiboles to be titaniferous richterites. The pleochroic scheme of the mineral is a = pale yellow; β = salmon pink to chocolate brown; γ = greenish yellow. Analyses of some amphiboles from this area (Table 3.10) show that they are deficient in Al and therefore Cundari considered that some Ti or Fe might occur in the four fold coordination. In contrast, the amphiboles from the Laacher See area, studied by Duda and Schminke (1978) are Al-rich (Table 3.9).

3.11 Melilite

Melilite is a solid solution of akermanite ($Ca_2MgSi_2O_7$), gehlenite ($Ca_2Al_2SiO_7$), and soda melilite ($CaNaAlSi_2O_7$). The first two phases form a complete series of solid solution with a minimum at 73 wt.% of akermanite (Osborn and Schairer 1941). In their exploratory work on the system $Ca_2MgSi_2O_7$–$CaNaAlSi_2O_7$, Schairer et al. (1965) found that nepheline does not appear in the compositional range between akermanite$_{100}$ and akermanite$_{56}$ soda melilite$_{44}$, but

Fig. 3.2. The kalsilite (Ks)-lar-
nite (La)-forsterite (Fo)-quartz
(Qz) tetrahedron showing the
relation of akermanite (Ak) to
the potassium-rich phases kalsilite (Ks), leucite (Lc), and sanidine (Sa). (After Yoder 1973)

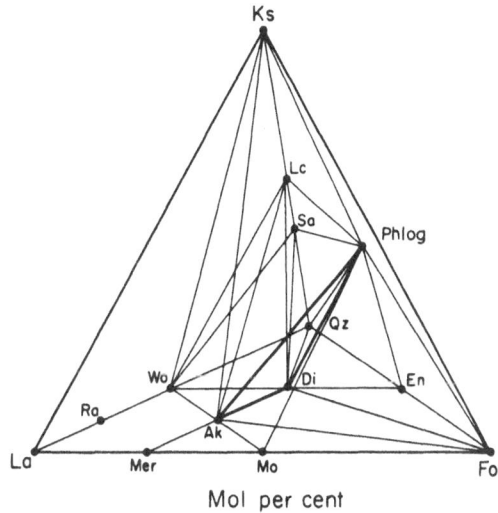

Mol per cent

3.12 Analcite

Association of leucite and analcite is common in many potassium-rich
volcanic rocks. In the Highwood Mountains area, Larsen (1940) noted
the occurrence of leucite and analcite-bearing phonolites. Weed and
Pirsson (1896) described a missourite (mica-bearing olivine leucitite)
in which secondary analcite is found in association with leucite. In
Totilto Park in Navajo, Williams (1936; p. 152) described a leucitite
dike in which analcite is present along with zeolites. The replacement
of leucite by analcite is frequently observed in the potassic rocks of
various areas of Italy. Analcite is also reported from the leucite-bearing
rocks of Utsuryo Island (Tsuboi 1920).

Analcite ($NaAlSi_2O_6 \cdot H_2O$) is cubic and, mineralogically and tex-
turally, is so similar to leucite (except for lack of multiple twinning)
that sometimes it has been misidentified as leucite. For example, the
alkalic rocks of Ryudo, Kankyo-Hokudo (Korea) were first described
as "leucite basalt" (Yamanari 1920) but later study by Tomita (1967)
revealed that the so-called leucite is analcite. Many so-called leucite
grains of the potassic rocks of Navajo petrographic province (Williams
1936) have now been established to be analcites (D. Smith personal
communication 1977). Analyses of two analcites from potassic rocks
are given in Table 3.10.

Experimental study of analcitization of leucite by Gupta and Fyfe
(1975) is discussed in Chap. 13.

3.13 Accessory Minerals

3.13.1 Sodalite. Minerals belonging to the sodalite group such as sodalite, nosean, and hauyne occur in association with nepheline or leucite-
bearing rocks, though their amount is small. They are all cubic with
large cavities in the frame work, which are occupied by Cl or SO_4
(Barth 1932). In sodalite, $Na_8(Al_6Si_6O_{24})Cl_2$, Na content is almost
constant, being only slightly substituted by both K and Ca. Nosean,
$Na_8(Al_6Si_6O_{24})SO_4$, has SO_4 radicals. They are usually pale blue in
color.

Tomisaka and Eugster (1968) synthesized sodalite, nosean, and other
related minerals by mixing appropriate chemicals and heating them in
sealed platinum or gold capsules, with or without water at temperatures
up to 1100 °C and pressures up to 2 kb. They showed that there is a
complete series of solid solution with a continuous change of physical
properties at temperatures higher than 1050 °C, but there is a wide
solvus between sodalite and nosean below this temperature.

Hauyne typically occurs in the potassic rocks of the Eifel (Duda
1975) and Vesuvius (Rittmann 1931). Nosean is often noted as an accessory mineral in the rocks of East Eifel (Rath 1864) and sodalite has
been described from the rocks of Utsuryo Island (Harumoto 1970).

3.13.2 Apatite. Apatite is a very commonly occurring accessory mineral
in leucite-bearing rocks of West Kimberley, Spain, the Leucite Hills
(Carmichael 1967), volcanic fields of Roccamonfina (Appleton 1972),
New South Wales (Cundari 1973) and Somma-Vesuvius (Savelli 1967).
The high phosphorous content of the rocks of various other localities
such as Java and Celebes, Navajo-Hopi, Manchuria, Tristan da Cunha
and Utsuryo Island (wt. % of P_2O_5 sometimes exceeding 1; Tables 5.4,
5.5, 5.13, and 5.16), suggests that this mineral should be present as a
common mineral in the rocks of these localities also.

Deer et al. (1963) and Cruft (1966) suggested various substitutions
that may take place in the apatite structure, of which the common re-
placements are Mn and Sr for Ca and rare earths for Ca. Coupled sub-
stitutions of Si^{6+} for P^{5+} along with Si^{4+} for P^{5+} or Na^+ for Ca^{2+} may
also occur (Carmichael 1967). The apatites from the leucite-bearing
rocks of southern Spain, the Leucite Hills and West Kimberley were
studied by Carmichael (1967), who noted that the rare earth concentrations of the apatites from these three areas fall in the following ranges
(in wt. %): La (0.2-0.6), Ce_2O_3 (0.5-1.0), Pr_2O_3 (0.2-0.4), and

Nd_2O_3 (0.2–0.5). He further observed that the West Kimberley apatites are notably higher in rare earth concentrations than the apatites from the other two areas.

3.13.3 Priderite. Because of high concentration of K, Ba, Fe, and Ti, priderite is a common accessory mineral in the rocks of the Leucite Hills (Carmichael 1967) and West Kimberley (Prider 1939). Its composition is $(K, Ba)_{2-y}$ $(Ti, Fe, Al)_{8-z}O_{16}$, where the value of y is close to 1, and that of z is very small.

3.13.4 Wadeite. A higher concentration of Zr is also reflected in the presence of wadeite $(Zr_2K_4Si_6O_{18})$ in some madupites from the Leucite Hills, and leucite-bearing lamproites from West Kimberley. Careful examination of the accessory minerals from other areas should also reveal the presence of priderite and wadeite.

3.13.5 Other Minerals. Melanite, perovskite, and rutile have been reported from all areas where leucite-bearing rocks are found. Magnetite, ilmenite, and chromium-bearing spinel are also common accessory minerals.

Chapter 4   Minor and Trace Element Geochemistry, Initial $^{87}Sr/^{86}Sr$ Ratios and Oxygen Isotopic Ratios of Leucite-Bearing Rocks

4.1 Trace Element Geochemistry

Holmes and Harwood (1937) studied the trace element geochemistry of leucite-bearing rocks of the Birunga volcanic province and found that these differ chemically from the granites and pegmatites in the abundance of Ti, Cl, P, Ba, and Sr and in the paucity of Li, B, and Be. They differ from basaltic rocks by their abundance of F, P, Ba, and Sr and presence of Zr. Higazy (1954) studied the trace element contents of the rocks of Birunga and Toro-Ankole and found that the notable chemical features of these rocks from the latter area are also the relative abundance of Sr, Ba, Rb, and Zr, which are uncommon in ultramafic rocks. They are high in Cr and Ni contents, except for the rocks poor

Table 4.1. Trace element composition of katungites. (After Higazy 1954)

Element	1	2	3	4	5	6	7
Rb	200	150	240	220	150	380	200
Li	10	12	8	7	6	6	8
Ba	2,800	2,600	7,000	2,000	1,800	4,500	2,900
Sr	7,500	3,800	10,000	7,500	4,000	9,500	4,500
Cr	700	800	290	650	500	1,200	900
Co	85	70	60	80	80	70	65
Ni	230	140	100	200	160	180	270
Zr	800	1,100	1,100	850	1,200	800	1,200
La	30	30	100	50	30	70	40
Y	30	30	30	30	30	30	30
Cu	80	60	200	70	60	50	70
V	320	350	170	260	350	210	250
Ga	25	30	25	25	30	20	25
Sn	–	–	5?	–	–	35	–
Pb	10	10	10	10	10	10	10
Se	10	–	–	10	10?		

1–3. Katungite from Katwe-Kikorongo, Uganda
4–6. Katungite from Bunyaruguru, Uganda
7. Katungite from Katunga, Uganda

in Mg, namely melilite leucitite and melilite-potash nephelinite. The other characteristic features include low Li content and relatively high La and Y (La > Y). Although the rocks of both areas have similar trace element contents, the Y content of the Birunga rocks is higher than the La content. The trace element contents of some of the leucite-bearing rocks are summarized in Tables 4.1 and 4.2 and Fig. 4.1.

Wade and Prider (1940) found that in contrast to basalts, the leucite-bearing rocks of West Kimberley have also a higher concentration of trace elements. However, they noted that BaO, in particular, is higher in the rocks of this province (0.89% to 1.00%), probably because the potassium content of this area is higher than that of other areas. SrO varies from 0.16% to 0.22%, where it is present in potash richterite with an average of 0.5% TiO_2 content of the rocks from West Kimberley is also high (4.08% to 7.39%), where it is mostly present in phlogopite and

Table 4.2. Trace element composition of some potassic rocks from Uganda. (After Higazy 1954)

Elements	1	2	3	4	5
Rb	300	150	160	180	140
Li	10	9	9	8	18
Ba	1	2,800	2,600	2,000	3,000
Sr	1	4,500	5,500	4,500	5,000
Cr	1	10	50	75	70
Co	70	60	50	40	50
Ni	30	35	30	35	40
Zr	1,100	900	700	650	750
La	80	30	30	30	30
Y	100	45	40	40	40
Cu	45	35	12	15	15
V	340	320	300	320	350
Ga	35	40	50	50	50
Pb	10	10	10	10	10
Mo	2	–	1?	1?	1?
Ag	1	–	1	–	–

1. Potash nepheline melilitite, Niragongo crater, Birunga, Uganda. New analysis by Imperial Chemical Industries Limited, Research Department, Billingham, Co. Durham
2. Kivite, lava on the northern slope of Nyamuragira, Birunga, Uganda. New analysis by the same company as indicated in 1
3. Leucite basanite, Kituru, new volcano south-south west of Nyamuragira, Birunga, Uganda. Anal. W.H. Herdsman
4. Leucite basanite, Muhubuli, Uganda. Anal. W.H. Herdsman
5. Leucite trachybasalt. Fissure north west of Kituru, Uganda. Anal. W.H. Herdsman

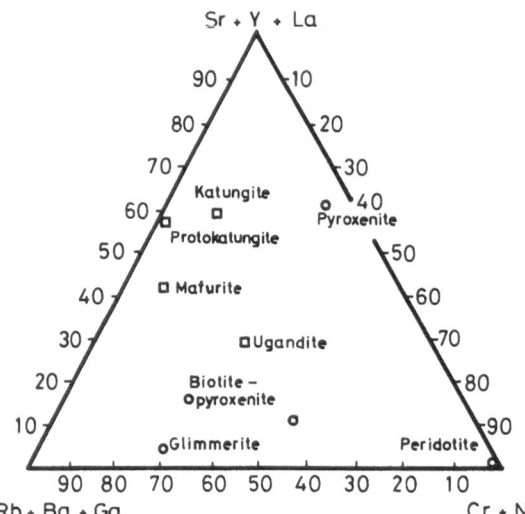

Fig. 4.1. Plot of the trace element contents of potassic rocks in a triangular diagram of (Sr + Y + La), (Rb + Ba + Ga) and (Cr + Ni). (After Higazy 1954)

amphibole. Sometimes it also occurs as rutile. The rocks of Ruwenzori (Holmes 1932; p. 420) and Java (Iddings and Morley 1915; Table 5.4) have also high TiO_2 although generally less than 4%.

Trace element abundances of leucite-bearing rocks from Spain and the Leucite Hills have been determined by Carmichael (1967), who found that the Y content of the jumillite lava from Spain is 40 ppm, whereas in the rocks of the Leucite Hills it is 25 ± 10 ppm. The La content of the jumillite is 200 ppm, whereas in some orendites its concentration is 130 ppm. In wyomingites it varies from 240 to 260 ppm and the highest concentration of 360 ppm is noted in a madupite. The Rb concentration of some of the potassic rocks is as follows: 170 ppm (jumillite; Mauricia, Spain), 460 ppm (wyomingite; Steamboat Springs, Leucite Hills), 290 ppm (orendite; North Table Mountain, Leucite Hills), and 205 ppm (madupite; west side of Pilot Butte, Leucite Hills). The values of K/Rb ratio in the specimens described above were found by Carmichael to fall within the range of basaltic rocks. This ratio is 250 in the case of jumillite and varies from 190 to 320 in the rocks from the Leucite Hills.

Trace element concentration of the rocks from Somma-Vesuvius was studied by Savelli (1967; Table 5.8), who observed that the leucitites from Vesuvius are characterized by higher concentration of F, Cl, Mn, Cu, Sr, and Ba than those of the tephrites of the Somma lavas. Vesuvius lavas also contain 30 times more Cl than the Somma lavas. Minor and trace element contents of the potassic rocks from Vico

(Cundari and Mattias 1974) and Roccamonfina (Appleton 1972) are summarized in Tables 5.7 and 5.9.

The leucitites of New South Wales have also high concentrations of Ti, Cr, Zr, Rb, Sr, and Ba (Table 5.3). Cundari (1973) found that relative to alkali basalt the mean composition of the leucitites is characterized by higher concentration of Cr and Sr by a factor of two, Rb by a factor of four. Only Co of the leucitites was comparable to that of alkali basalt. Trace element compositions of the Eifel rocks are summarized in Table 5.13.

Baker et al. (1964) noted that in the crystallization sequence of lavas from the Tristan da Cunha islands there is a gradual decrease in the V content of the rocks, produced at the later stages. However, in such a sequence the concentration of Ba (700–800 ppm), Rb (100 ppm) and Zr (300 ppm) increased respectively to 1300, 400, and 500 ppm in the rocks formed during the final stages of crystallization. Sr contents are 700 ppm in the trachybasalts and trachyandesites, however, in the lavas differentiated at a late stage, its concentration is as low as 10–20 ppm. Ni content of the rocks of Tristan da Cunha was found by them to be very low.

The Rb contents of the potassic rocks from various localities are summarized in Table 4.3.

4.2 The Initial ^{87}Sr/^{86}Sr Ratios of Highly Potassic Lavas and Other Rock Types

Hoefs and Wedepohl (1968) found that the ^{87}Sr/^{86}Sr ratios of the Tertiary tholeiites, alkali olivine basalts, and olivine nephelinites from lower Saxony and Hessia Provinces and the leucite-nepheline tephrites from the Laacher See area, have values ranging between 0.7031 and 0.7054, which are quite similar to those of the Hawaiian rocks. The ratios in the leucite-nepheline tephrites from Niedermendig and Ettringer Feld are 0.7049 and 0.7046 respectively, whereas those of a nepheline leucitite from Hochsimmer (0.7052) and a phonolite from Schell Kopf (0.7093) are slightly higher. All these localities are from the East Eifel region of West Germany. Hoefs and Wedepohl also calculated the ^{87}Sr/^{86}Sr ratios of the rocks from Somma-Vesuvius and Phlegrean Fields (Table 4.3).

Barbieri et al. (1975) studied the ^{87}Sr/^{86}Sr ratios of the minerals of some ejected blocks found in the Albano "pepirano", one of the

Table 4.3. $^{87}Sr/^{86}Sr$ and Rb/Sr ratios and Rb contents of potassic rocks

Locality: rock type	Initial $^{87}Sr/^{86}Sr$	Rb/Sr	Rb (ppm)	Sources
Western Rift, Africa				
Absarokite	0.7076	0.141	138	1
Basanite	0.7043	0.104	127	1
Katungite	0.7047	0.052	124	1
Kivite	0.7067	0.107	111	1
Leucitite	0.7058	0.102	152	1
Mafurite	0.7050	0.059	129	1
West Kimberley, Australia				
Cedricite	0.7125	0.463	513	2
Mamilite	0.7210	0.228	252	2
Wolgidite	0.7127	0.349	596	2
Fitzroyite	0.7127	0.358	644	2
Jumilla, Spain				
Jumillite	0.7136	0.058	119	2
Leucite Hills, USA				
Wyomingite	0.7062	0.103	296	2
Orendite	0.7060	0.133	277	2
Madupite	0.7057	0.065	218	2
Navajo Province, USA				
Trachy basalt	0.7078	0.091	105	2
Minette	0.7075	0.047	74.5	2
Somma-Vesuvius, Italy				
Leucite-bearing tephritic phonolite	0.7102			3
Phlegrean Fields, Italy				
Phonolite	0.7089			3
Trachyte	0.7073			3
Alban Hills, Italy				
Melilite leucitite	0.7110			4
Leucititic pyroclastic rock	0.7110			4

(1) Bell and Powell 1969, (2) Powell and Bell 1970, (3) Hoefs and Wedepohl 1968, (4) Barbieri et al. 1975

largest pyroclastic formation of the Alban Hills volcanic district (Roman Comagmatic region). They also determined this ratio for one leucititic lava (Table 4.3). Hurley et al. (1966) and Volmer (1976) found that the Sr isotopic ratios of leucitites from the Alban Hills lie between 0.7095 and 0.7114. Barbieri et al. (1975) compared the Sr isotopic data of leucite-bearing rocks from various areas with respect to their Rb contents (Fig. 4.2).

The initial ^{87}Sr/^{86}Sr ratios of potassic rocks of Birunga and Toro-Ankole vary significantly and range between 0.7036 and 0.7111 (Bell and Powell 1969). The Sr isotopic ratios of some of the leucite-bearing rocks of the Western Rift region are given in Table 4.3. The feldspar-bearing lavas of this area have higher ratios than the feldspathoid-bearing varieties (Table 4.3). For example the Sr isotopic ratios of absarokite (0.7076) and kivite (0.7067) are higher than those of katungite (0.7047), leucitite (0.7058), and mafurite (0.7050). The initial ^{87}Sr/^{86}Sr ratios of the potassic rocks of the Birunga and Toro-Ankole regions have a positive linear correlation with the Rb/Sr ratio (Fig. 4.2) and a negative correlation with Sr, Nb, and Zr abundances.

The Sr isotopic ratios of the potassic rocks from Spain, Australia, and the United States, as determined by Powell and Bell (1970) are listed in Table 4.3, which shows that the rocks of West Kimberley and

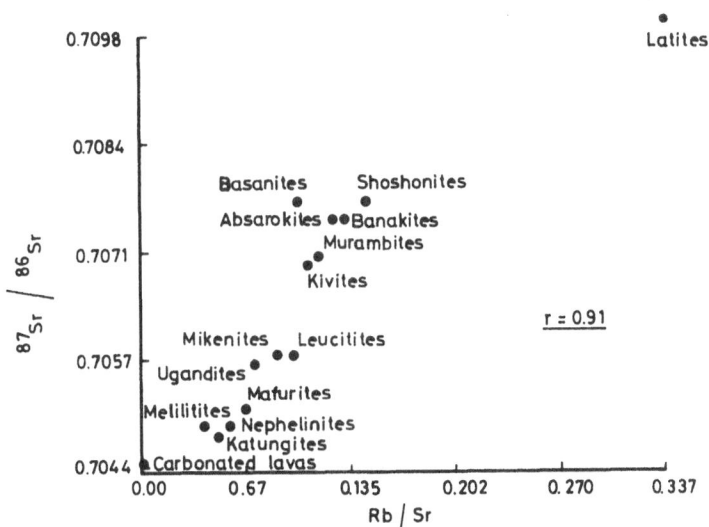

Fig. 4.2. Plot of ^{87}Sr/^{86}Sr ratio vs Rb/Sr ratio (wt. %) using the average values for the different rock types. (After Bell and Powell 1969)

Jumilla have the highest Sr isotopic ratios, comparable to those of the average crustal materials.

4.3 $\delta^{18}O$-contents of Leucite-Bearing Rocks

$\delta^{18}O$ was defined by Taylor and Epstein (1963) as follows:

$$\delta^{18}O = \left[\frac{(^{18}O/^{16}O)_{sample}}{(^{18}O/^{16}O)_{oceanic\ water}} - 1 \right] \times 100 \ .$$

Garlick (1966) calculated the ^{18}O values of the mineral specimens of leucite-bearing lavas of Lake Kivu (Zaire, Africa), leucite monchiquites of Bohemia, leucite phenocrysts of Roccamonfina lava and leucite basanites of Bosco Trecase, Vesuvius (Italy). The ^{18}O values of Garlick are summarized in Table 4.4. The high ^{18}O contents of lavas from Vesuvius were considered by him to be due to contamination with dolomitic rocks underlying the volcano.

Turi and Taylor (1976) measured the $^{18}O/^{16}O$ ratios of leucite-bearing lavas, pyroclastics, and other related rocks from the Roman Province of Italy. They observed a general increase northwards in the ^{18}O values in the following sequence: Ischia (5.8 to 7.0), Somma-Vesuvius and Phlegrean Fields (7.3 to 8.3), Alban Hills (7.3 to 8.7), Mount Sabatini

Table 4.4. $\delta^{18}O$ values of some leucites and augites from potassic rocks of various areas. (After Garlick 1966)

Rock type	Locality	Phase	$\delta^{18}O$	Average deviation
Leucititic lava	Lake Kivu, Zaire	Leucite	+6.8	±0.20
		Augite	+5.9	–
Leucite monchiquite	Boehemia	Augite	+5.3	±0.02
Leucite phenocryst	Roccamonfina	1	+6.91	–
		2	+6.93	–
		3	+6.96	–
Leucite basanite	1760 flow, Uncino, Vesuvius	Leucite	+7.7	–
		Augite	+6.9	–
Leucite tephrite	1906 flow, Bosco Trecase, Vesuvius	Leucite	+8.2	±0.17
		Augite	+7.2	–

1-3 are analyses of core, intermediate portion and rim of the leucite phenocryst

Table 4.5. $\delta^{18}O$-contents of the potassic rocks from Italy and Australia. (After Turi and Taylor 1976)

Rock type	$\delta^{18}O$ (0/00)			Locality
	Whole rock	Leucite	Other minerals	
Nepheline leucitite	8.63 ±0.002	–	25.7[a]	Colli Albani
Nepheline-melilite leucitite	8.73	–	26.1[a]	Colli Albani
Melilite leucitite	8.33 ±0.005	7.75 ±0.15	26.2[a]	Colli Albani
Tephritic leucitite	7.32 ±0.10	–	6.54[b]	Colli Albani
Leucite inclusion in lava	7.39	7.86		Colli Albani
Biotite inclusion in a tuff			7.61[c]	Villa Senni
Tephritic leucite phonolite	9.67	9.62		Monti Sabatini
Leucitite	7.30	7.68		Monti Sabatini
Leucite phonolite	8.63			Monti Sabatini
Tephritic phonolitic tuff		7.65 ±0.002	7.70[d] ± 0.002	Monti Sabatini
Leucite phonolite	9.4	9.18		Vico Volcano
Tephritic leucite phonolite	8.91	8.28		Vico Volcano
Tephritic leucite phonolite	8.40	8.25	7.08[b]	Vico Volcano
Leucite tephrite	8.64			Monti Vulsini
Leucite tephrite	9.78 ±0.14		9.26[d]	Monti Vulsini
Phonolitic olivine-leucite tephrite	9.93 ±0.13	9.84		Monti Vulsini
1944 flow of Vesuvius	8.02	7.93 ±0.15	–	Somma-Vesuvius
Porphyritic leucite tephrite	8.29	–		1804 lava flow, 3 km north of Annunziata
Trachytic tuff			8.40[d]	Phlegrean Fields
Leucitite[e]	7.35 ±0.14			Begargo Hills, New South Wales (Australia)

[a-d] indicate analyses of calcite, augite, biotite, and sanidine respectively

[e] Only locality outside Italy

(7.3 to 9.7), Vico Volcano (7.4 to 10.2), Monti Vulsini (8.7 to 11.7). The Sr isotopic ratios of the rocks of the Roman Province also show a similar northward increase. A marked increase in the ^{18}O value was noted by them just north of Rome, where the rocks of the Roman Province just begin to overlap Tuscan calc-alkaic rocks. This observation led them to believe that the observed increase in the ^{18}O contents and $^{87}Sr/^{86}Sr$ ratios might either be due to direct mixing of the parent magma with the Tuscan rocks, which are high in the ^{18}O content, or to the fact that the high ^{18}O country rocks underwent heating during 2 million years of Tuscan igneous activity. They estimated that the strongly undersaturated magma, which was probably derived from the upper mantle, had a $\delta^{18}O$ value of +6 and Sr isotopic ratio of 0.704 to 0.705 and SiO_2 content of less than 44%. The ^{18}O contents of the leucite-bearing rocks from various localities of Italy and New South Wales are given in Table 4.5.

Chapter 5 Distribution of Leucite-Bearing Rocks

Leucite-bearing rocks are found in widely scattered localities all over the world (Fig. 5.1). The best known localities include volcanic fields of east and southeast Ruwenzori, Uganda; Birunga volcanic province of central Africa; West Kimberley, and New South Wales of Australia; Java and Celebes of Indonesia; the Leucite Hills of Wyoming; the Massif Central region of France, the Roman Province of Italy; Mauricia and Almaria provinces of Spain and the Laacher See district of W. Germany. In addition to these classical localities, Gittins (personal communication, 1977) has reported the occurrence of leucite-bearing rocks from the east coast of Greenland, and Baker and Rea (1978) have described leucitic rocks from Huopi, Patagonian Plateau, which is located between the Southern Andes and Lake Colhue. It should be noted here that all these localities are confined to continental regions.

Occurrence of leucite-bearing rocks in oceanic areas is very limited. Five localities have been reported so far. They are: Cape Verde Island, the Kerguelen Archipelago, Utsuryo Island, Marquesas Island, and Tristan da Cunha Islands. However, Baker et al. (1964) consider that of all these five localities, Tristan da Cunha is the only example of leucite-bearing rocks from a truly oceanic island.

There are also reported occurrences of leucite-bearing rocks from several other localities, which include Sardinia, Italy; St. Paul Island; Trebizond, Turkey; Lake Urumiah, Iran; Ampasindava, Madagascar, and Gaussberg, Antarctica (Iddings 1913). However, as more recent petrologic studies are lacking these will not be discussed.

A description of the important localities of highly potassic petrographic provinces is given in the following sections. Rocks which contain only pseudoleucite but not leucite will be discussed in Chap. 13, together with the genesis of pseudoleucite.

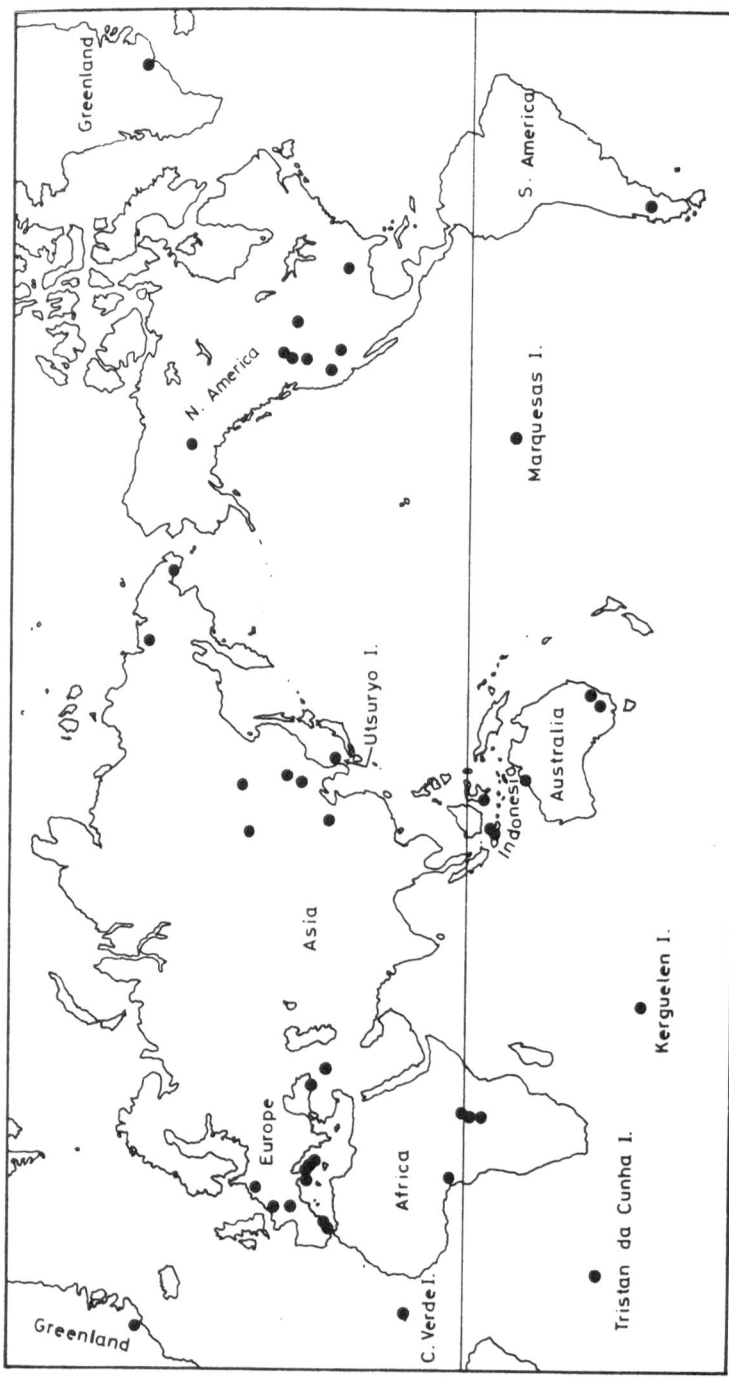

Fig. 5.1. Plot of different localities of leucite-bearing rocks in a world map

5.1 Volcanic Fields, East and Southeast of Ruwenzori, Uganda

Bright (1909; cited by Holmes and Harwood 1937) described potassium-rich volcanic rocks from Kichwamba, Kawata, Kyawata, and the Fort Portal area near Ruwenzori. He noted the remains of breccias, numerous tuff-cones together with circular vents, drilled by explosive gaseous eruption, and largely depressed areas of subsidence. The rocks occur as volcanic ejecta, consisting almost entirely of tuffs and agglomerates. From their positions these volcanic systems appear to be connected with the fractures bounding the Somliki Valley, which probably resulted by trough subsidence or a rift between roughly parallel faults, the subsided area being now occupied by Lake George. Late Pleistocene structural movements stimulated an explosive phase of volcanic activity in Post-Kaiso times. In all the volcanic areas of Kichwambe craters, lavas were blown through crystalline basements and through the Kaiso sediments and tuffs. No continuous lava flows have yet been found in this area. However, near Fort Portal, tuffs often locally grade into highly vesicular lavas. The earlier tuffs are mostly melilite basalts, which according to Holmes and Harwood (1932, p. 375) contain fragments and blocks of biotite pyroxenite and related rock types, which constitute cognate subvolcanic rocks. The melilite basaltic tuffs are succeeded by rocks transitional to leucitite (near Fort Portal) and a highly carbonated melilite-nepheline leucitite. The cognate ejected blocks also include melanocratic varieties of potassium-rich nephelinites and leucitites, which were described as potash ankaratrite. The general setting of the localities of the potassium-rich rocks of the Rift Valley System of Central and Equatorial Africa is shown in Fig. 5.2.

Holmes (1950) also described the volcanic fields near Ruwenzori, and those adjoining the Western Rift of Central Africa. He noted the presence of several areas of tuffs and explosion craters accompanied in some areas by lava flows extending from near surface plugs and sheets. The following fields were recognized by Holmes (Fig. 5.2): (1) Katunga, in the plateau country in the southern part of the province, (2) Bunyaruguru, (3) Katwe-Kikorongo in the rift zone between Lake Edward and Lake George, (4) Bukangara, on the slopes, leading up to the southern foot of Ruwenzori, possibly an extension of (3) in the north west, (5) an isolated unknown explosion crater, lying to the west of Ruwenzori, (6) a line of craters of Mohokya, (7) Ndale, where the low wall of the Lake George depression dies to the north, (8) Fort Portal, where the eastern wall of the Lake Albert Rift dies out to the south,

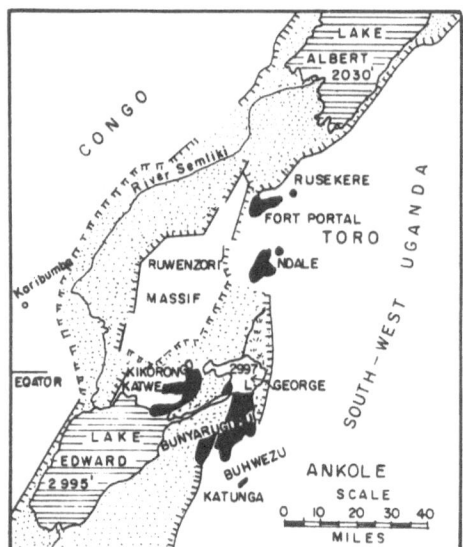

Fig. 5.2. Relationship of the volcanic fields of Toro-Ankole to Ruwenzori and Western Rift Valley System. (After Holmes 1950)

and (9) Rusekere, on the northeastern plateau. Most of the rocks occurring in this area are characterized by ultramafic potassic rocks, which are composed of various combinations of melilite, augite, kalsilite, leucite, and potassium-rich glass. The rock types include katungite, ugandite, and mafurite, with all possible transitions between them. Volcanic activity was explosive and the pyroclasts contain fragments of both basement rocks and of a subvolcanic suite composed of pyroxene and/or biotite and/or olivine, which was called the O.B.P. series by Holmes.

5.2 Birunga Volcanic Province

Holmes and Harwood (1937) made extensive studies of the rocks of Birunga. The area is situated at the junction of Zaire, the southwestern part of Uganda, and the northwestern part of Rwanda (Fig. 5.3). Bufumbira forms a part of the greater volcanic fields of Birunga, which extend along, and to some extent across, the trend of Rift Valley depression for more than 80 km north and north west of Lake Kivu. The rift movements in western Uganda probably started at about the beginning of Miocene with the initiation of the Albertine Depression and the Ruwenzori uplift. Repeated earthquakes in the area show that the movements have continued at intervals up to the present time. Accord-

Fig. 5.3. Major volca-
noes of the Birunga
volcanic field. (After
Holmes 1965)

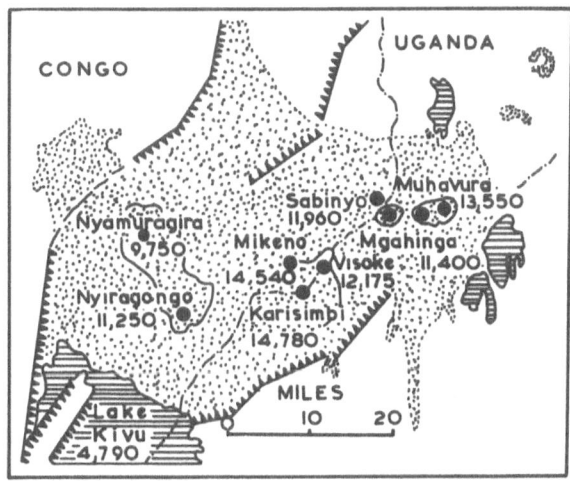

ing to Bell and Powell (1969), the Birunga field is composed of eight major volcanoes with four hundred subsidiary cones. The well-known active volcanoes include Nyamlagira in the Birunga province of Zaire. According to Combe (1933) the volcanic activity in the area probably started in the early Pliocene with the formation of the Sabinyo and Mikeno volcanoes, whereas the other volcanoes such as Mahavura, Magahinga, Karsimbi, and Visoke probably erupted during the Middle and Late Pleistocene.

Holmes and Harwood classified the rocks of Birunga into six petro-graphic series:

1. An olivine leucitite series, related to the dunite-mica pyroxenite series.
2. A melilite feldspathoidal series of rocks, developed between Nina-gongo and Lake Kivu.
3. A leucite basanite series from murambite through kivite to leucite basanite.
4. A potash trachybasalt series related to absarokite-shoshonite-bana-kite series.
5. A potash trachyandesite or latite series, represented by the lavas of Sabinyo.
6. A limburgite trachybasalt series in which the K_2O/Na_2O ratios are < 1.

Chemical analyses of some potassic lavas of the area are given in Table 5.1.

Table 5.1. Analyses of leucite-bearing rocks from central and equatorial Africa

	1	2	3	4	5	6	7
SiO_2	44.21	44.24	45.90	48.05	33.52	45.90	44.42
TiO_2	2.31	3.22	2.80	2.96	6.04	3.15	2.86
Al_2O_3	15.03	13.31	15.33	13.42	8.04	16.20	17.09
Fe_2O_3	3.79	5.65	1.85	3.54	5.88	1.87	3.11
FeO	7.53	5.29	9.63	7.68	5.50	10.35	7.26
MnO	0.18	–	0.03	–	0.15	0.19	0.20
MgO	6.61	5.68	7.30	6.41	13.54	4.02	3.76
CaO	10.87	12.72	9.83	9.05	15.22	11.12	8.47
Na_2O	3.75	4.06	2.62	3.06	1.42	3.19	4.34
K_2O	4.07	4.71	3.28	3.85	4.26	3.75	7.29
H_2O^+	0.36	0.95	0.37	1.55	2.34		
H_2O^-	–	–	0.09	–	1.68	–	–
P_2O_5	0.90	0.64	0.66	0.58	0.82	0.26	1.18
BaO	0.17	–	0.12	–	0.15	–	–
SrO	0.09	–	0.05	–	0.44		
CO_2	0.02	–	–	–	0.96	–	–
Cl	0.11	–	0.06	–	–	–	–
F	0.11	–	0.06	–	–	–	–
S	0.02	–	0.01	–	–	–	–
Cr_2O_3	0.03	–	0.02	–	–	–	–
Total	100.16	100.47	100.01	100.15	99.96	100.00	99.98

1. Olivine leucitite, Lutale area (Holmes and Harwood 1937; p. 75)
2. Olivine-nepheline leucitite, Kitale (Holmes and Harwood 1937; p. 97)
3. Kivite, Busamba (Holmes and Harwood 1937; p. 104)
4. Leucite basanite, Muhavura (Finckh 1912; p. 18)
5. Kalsilite katungite, lapilli separated from a tuff, Bunyaruguru (Holmes 1950; p. 780)
6. Leucite tephrite from Kivu, Nyiragongo (Denaeyer 1965; quoted in Sahama 1973)
7. Tephritic leucitite from Kivu, Nyiragongo (Denaeyer 1965; quoted in Sahama 1973)

Ferguson and Cundari (1975) also studied the leucite-bearing lavas from the volcanoes of Mikeno, Magahinga, and Mahavura and associated volcanic cones of the Bufumbira region. They plotted the mineralogical compositions of these rocks on an A–P–F diagram (A = alkali feldspar, P = plagioclase, F = feldspathoid) as shown in Fig. 5.4. The diagram shows that the rocks of the Bufumbira area constitute two petrographic series, both of which diverge from a common melabasanite of composition A. The first series (B) contains the following rock types with decreasing color index: leucite tephrite, tephritic leucitite, leucitite, and phonolitic leucitite. Point C corresponds to the composition of a glass,

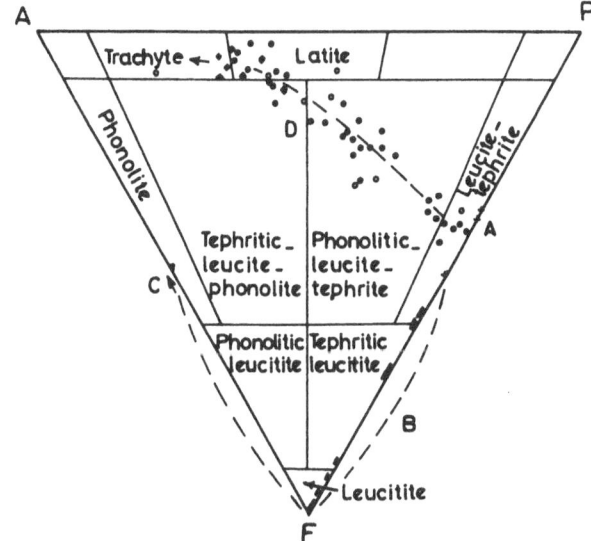

Fig. 5.4. *A* (alkali feldspar) – *P* (plagioclase) – *F* (feldspathoid) diagram of Birunga rocks. (After Ferguson and Cundari 1975)

designated as a (leucite) phonolite. This series was called the leucitite series. The other series (D) called the phonolitic tephrite series, consisted of the following rock types in order of decreasing color index: leucite tephrite, phonolitic leucite tephrite, tephritic leucite phonolite, latite, and trachyte. The phonolitic tephrites typically contain xenoliths, consisting of the following assemblages: clinopyroxene-olivine-biotite (without ulvospinel). Xenocrysts of biotite and kaersutite, rimmed with reaction corona of Fe-oxides are also found in the phonolitic tephrites.

On the basis of their studies the following conclusions were obtained:

a) The Bufumbira phonolite tephrite series apparently resembles the sodic mildly potassic series of St. Helena and Gough islands.
b) The shoshonitic rocks of Papua, New Guinea evolved into the silica-saturation series in contrast to the Bufumbira phonolitic series.
c) The Bufumbira leucitite series is similar to the Shonkin Sag laccolith and the Alban Hills lava in their compositional trends.

d) The Bufumbira basanites represent the most primitive leucite-bearing lavas, similar in composition to those found in the Toro-Ankole
area to the north of the Bufumbira area.

Ferguson and Cundari believed that the leucitite series probably
formed by fractionation of an olivine and biotite-bearing pyroxenite
assemblage derived from an ultramafic magma. At pressures of 1 to 2 kb,
biotite becomes unstable and the liquid becomes enriched in K_2O. They
thought that the phonolitic tephrite series may have formed by fractionation of a biotite pyroxenite assemblage from a parental magma of
leucite tephrite composition in the deeper crust (2 to 6 kb). The presence of biotite and kaersutite xenocrysts in the lavas belonging to the
series possibly indicates that they were formed under similar conditions
as the rocks of Tristan da Cunha, where kaersutite-bearing nodules have
been reported by Le Maitre (1969).

The most notable active volcanoes of the Birunga volcanic field are
Mt. Nyamuragira (Nyamlagira) and Mt. Nyiragongo (Niligongo, Niragongo), which also include other satellite eruption centers (Fig. 5.3).
Volcanism of this area was studied by Meyer (1955), Tazieff (1966),
and Sahama (1973). The chemical compositions of the rocks of the
two volcanoes were plotted by Sahama (1973) in an alkali feldspar-
plagioclase-feldspathoid-quartz diagram (Fig. 5.5).

In the rocks of Nyamuragira, plagioclase (labradorite to bytownite)
and titanaugite are the predominant crystalline constituents with variable amounts of olivine, small amounts of leucite, and nepheline. Ortho-
pyroxenes are associated with clinopyroxene in some rocks (Egoroff
1965).

The rocks of Nyiragongo have variable mineral compositions and
most of the rocks are feldspar-free pure leucitites or nephelinites with
either of these two feldspathoids predominating (Sahama 1973). In
some rock types melilites are abundant. According to Sahama titanaugite are common in leucitites and melilite-poor nephelinites, whereas
Al-poor salites and on some rare occasions acmitic pyroxenes are common in the melilite-rich nephelinites. Olivine occurs in minor amounts,
but in some rocks it is abundant. The most commonly found xenoliths
in these rocks are fragments of granitic basement rocks. Occurrence of
kalsilite in most melilite nephelinites is the regional characteristic of
this area (Sahama 1973). Chemical analyses of two leucite-bearing rocks
from Kivu, Nyiragongo are given in Table 5.1.

The Toro-Ankole area, located about 125–155 km north-northeast
of the Bufumbira field (Fig. 5.2), has been described by Holmes and

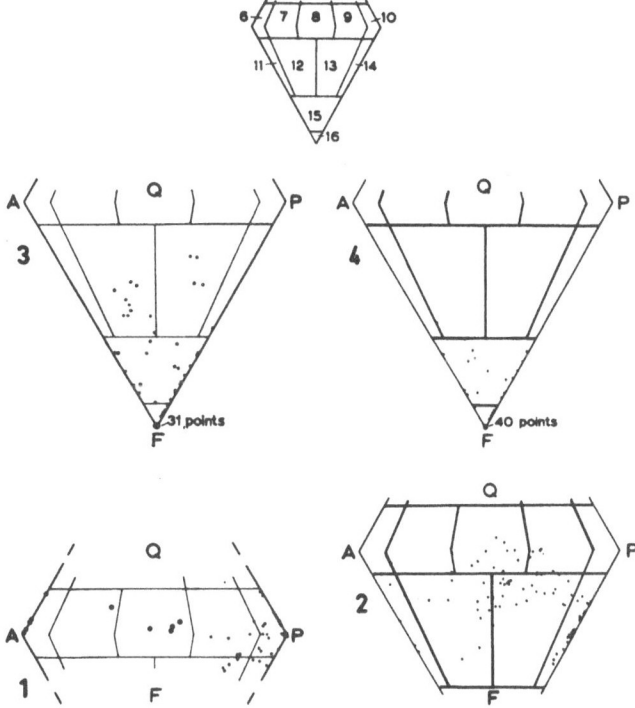

Fig. 5.5. R-normative variation of some Western Rift volcanics. Normative calculation is based on Rittmann's method. *Q* quartz; *A* alkali feldspar; *P* plagioclase; *F* feldspathoid. 1 South Kivu trchytes and basaltic rocks; 2 rocks of Nyamuragira; 3 rocks of the lava plain of the Nyiragongo area; 4 Nyiragongo area rocks of the main cone complex. (After Sahama 1973)

Harwood (1932), Holmes (1937), Combe and Holmes (1945), and Brown (1971). The lavas of the area belong to the olivine leucitite and olivine melilitite family and are completely devoid of feldspars. Katungites are also fairly common. The lavas often include xenoliths of phlogopite augite peridotite, and glimmerite. Fragments of partially fused granites are also present as xenoliths, the glassy portion of which chemically corresponds to almost pure leucite. Chemical analyses of the leucite-bearing rocks from the Western Rift are summarized in Table 5.1.

5.3 Leucite-Bearing Rocks of West Kimberley, Australia

Pseudoleucite and leucite-bearing rocks from the West Kimberley region of Western Australia, where they occur in volcanic craters, plugs, and fissure intrusions have been described by Wade and Prider (1940). The following nineteen localities (Fig. 5.6) were reported by them: (1) Kalaida Hills, (2) Mt. Abbott, (3) Oscar Plug, (4) Hills Cone, (5) Noonkanbah Hill, (6) Mamilu Hill, (7) Mt. Ibis, (8) Machell's Pyramid, (9) "P" Hills, (10) Wolgidee Hills, (11) Howe's Hill, (12) Mt. Gytha, (13) Mt. Cedric, (14) Moulamen Hill, (15) Bruton's Hill, (16) Mt. North, (17) Dadja Hill, (18) Fishery Hill, and (19) White Rocks. Figure 5.6 shows that a greater number of the occurrences are grouped in a synclinal depression between the Mt. Wynne anticline and the St. George Range anticlinal folds. Wade and Prider found that these rocks occur along two lines trending in two directions: north-northeast to south-southwest and west-northwest to east-southeast. These directions represent important faults and structural lines in Western Australia. The strike of basement Precambrian rocks in the eastern parts of the Kimberley division is north-northeast and south-southwest. The structure and distribution of the volcanic masses indicate that they ascended along fault planes, connected with the structures of the underlying Precambrian rocks. In every instance ultrapotassic magma is found to have intruded into the Permian sandstone, limestone, and shale. No significant thermal effects have been found in the limestone around the intrusive rocks. Some shale xenoliths occurring near Moulamen Hill, however, show the formation of abundant muscovite in their marginal parts, whereas the central parts remain as dark shale. From these observations, Wade and Prider concluded that the leucite-bearing lamproites in West Kimberley intruded as a crystal mush, giving very weak thermal effects.

The predominant rock types of the West Kimberley area are: fitzroyite, cedricite, and mamilite. Wolgidite is the most predominant rock type at Moulamen Hill, Mt. North, and Wolgidee Hill. Fitzroyite is the most common rock type found along the Fitzroy River basin, where it often includes quartz amygdules. Cedricite and mamilite are common rock types in the Mt. Cedric and Mamilu Hills, respectively. There are also some other rock types which are transitional in character between cedricite and fitzroyite, being similar to the wyomingite of the Leucite Hills. Some of the transitional rocks are found in "P" Hills.

Analyses of the rocks show the following features: (1) High K_2O content (as high as 12.60%) and its dominance over aluminum, (2) very low Na_2O content, (3) very high TiO_2 content (4.08% to 7.34%), (4) high concentration of BaO (as high as 1%), and SrO (0.16% to 0.22%), and (5) high MgO content (Table 5.2). The high concentration of K_2O is reflected by the abundance of leucite and the occurrence of phlog-

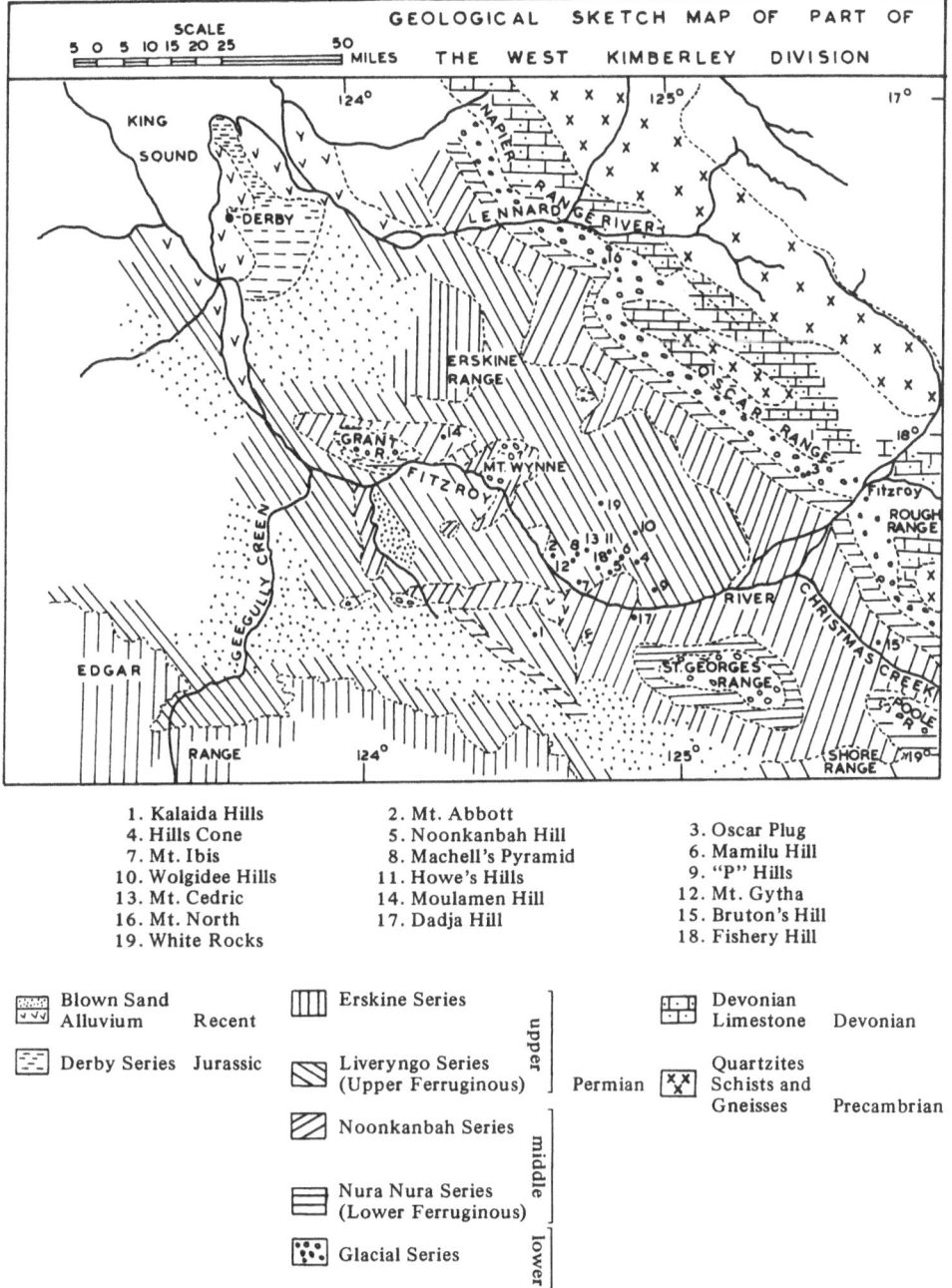

1. Kalaida Hills
4. Hills Cone
7. Mt. Ibis
10. Wolgidee Hills
13. Mt. Cedric
16. Mt. North
19. White Rocks

2. Mt. Abbott
5. Noonkanbah Hill
8. Machell's Pyramid
11. Howe's Hills
14. Moulamen Hill
17. Dadja Hill

3. Oscar Plug
6. Mamilu Hill
9. "P" Hills
12. Mt. Gytha
15. Bruton's Hill
18. Fishery Hill

Blown Sand	Erskine Series	Devonian Limestone Devonian
Alluvium Recent		
Derby Series Jurassic	Liveryngo Series (Upper Ferruginous)] upper Permian	Quartzites Schists and Gneisses Precambrian
	Noonkanbah Series	
	Nura Nura Series (Lower Ferruginous) middle	
	Glacial Series lower	

Fig. 5.6. Geological sketch map of part of the West Kimberley Division. (After Wade and Prider 1940)

Table 5.2. Analyses of leucite-bearing rocks from West Kimberley, Australia and Bearpaw Mountains, Montana, USA

	1	2	3	4	5	6	7
SiO_2	36.02	45.82	51.19	52.45	54.09	54.48	46.51
TiO_2	5.31	7.34	4.89	5.85	4.08	5.57	0.83
Al_2O_3	5.32	6.86	8.53	8.64	11.67	9.87	11.86
Fe_2O_3	5.37	6.07	6.12	5.48	4.91	4.89	7.59
FeO	0.89	1.98	1.38	0.94	2.14	1.70	4.39
MnO	0.04	0.10	0.06	0.13	Trace	0.09	0.22
MgO	7.79	10.90	7.15	6.42	4.76	5.35	4.73
CaO	15.12	4.70	5.82	2.01	1.91	1.89	7.41
Na_2O	0.16	0.84	0.58	0.38	Trace	0.88	2.39
K_2O	7.25	8.82	9.02	10.42	12.60	11.06	8.71
H_2O^+	1.92	0.75	1.99	1.99	2.30	1.36	2.45
H_2O^-	0.90	2.40	1.26	2.89	0.73	0.89	1.10
P_2O_5	1.15	1.83	0.79	1.58	0.26	0.40	0.80
BaO	1.58	1.27	0.60	1.19	0.74	0.64	0.50
SrO	0.22	–	–	–	–	0.16	0.16
ZrO_2	0.19	–	–	–	–	0.07	–
FeS_2	0.21	0.07	0.22	–	–	0.20	–
Cr_2O_3	Trace	0.07	Trace	–	–	Trace	–
CO_2	9.84	0.08	–	–	–	-	–
SO_3	0.27	–	0.11	–	–	0.10	0.05
Total	99.72	99.93	99.71	100.37	100.19	99.60	99.70

1. Wolgidite. It also includes 0.17% Cr_2O_3
2. Wolgidite. It also includes 0.03 % V_2O_3
3. Cedricite
4. Fitzroyite
5. Diopside-magnophorite-phlogopite-leucite rock
6. Mamilite. It also includes 0.09% F
7. Leucitite. It also includes 0.4% Cl and 0.4% NiO

1–6 are analyses of rocks from West Kimberley, Australia (Wade and Prider 1940).
7 is an analysis of a rock from Bearpaw Mountains, Montana (Weed and Pirsson 1896)

opite and potash richterite. Because of a high concentration of Zr and Ti, wadeite, perovskite, and priderite are common accessory minerals.

5.4 Leucite-Bearing Rocks of New South Wales, Australia

Cundari (1973) made a detailed petrological and geochemical study of the leucite-bearing rocks from the central portion of New South Wales. Several outcrops had already been described by Judd (1887), Curran

(1891), Stonier (1893), Browne (1933), and Griffin (1957). The outcrops are generally arranged along the regional structural patterns of
north-northeast trend of the Paleozoic basement rocks. These localities
and the areas of exposure are as follows: Byrock (1 km^2), El Capitan
(6 km^2), Condoblin (2 km^2), Bygalorie (46 km^2), Tullibigeal (43 km^2),
Lake Cargelligo (12 km^2), Begargo Hill (2 km^2), Flagstaff Hill (2 km^2).
In addition to lava flows, pyroclastic materials are also found at Flagstaff Hill and Tullibigeal, where scoria cones are developed. The volcanic activity in these localities occurred through small vents, which were
buried by eruption of lavas in the Late Cenozoic time.

The lavas in these localities essentially contain various amounts of
diopsidic pyroxene, leucite, olivine, and iron-titanium oxide minerals,
with small amounts of titaniferous phlogopite and amphibole (richterite). Common accessories include alkali feldspar, nepheline, and
apatite. Chemical analyses of some of the leucitites from New South
Wales are given in Table 5.3. Cundari (1973) found that the leucitic
suite in general is similar to Bufumbira ugandites with respect to their
alkali and silica content and their MgO content; and high K_2O/Na_2O
ratios are similar to micaceous kimberlites. Table 5.3 shows that the
leucite-bearing rocks of New South Wales are characterized by high
Mg, K, P, and H_2O and low Si, Al, and Na. The crystallization temperature of the magma was considered by Cundari to have been above
700 °C and pressures lower than 2 kb.

5.5 Leucite-Bearing Rocks of Indonesia

Iddings and Morley (1915), Van Padang (1951), and Rittmann (1951)
studied the volcanic rocks of Indonesia. Leucite-bearing rocks in particular occur in the extinct volcano of Mt. Mouriah, northeast of Sema-
rang, on the northeast of Java. Associated with this are two other small
volcanoes: Paliaian and Tülering. Other leucitic lavas also occur in a
small volcano, Lorous, and a larger much eroded volcano, Ringgit, in
Besouki, Eastern Java. Leucite-bearing rocks in association with phonolite also occur in the Bawean Island between Java and Borneo. Mt.
Mouriah is largely made up of tuffs and breccia with small massive lava
flows of mainly leucite tephrites and leucitites, with a few olivine-bearing varieties. These rocks are of late Tertiary age. Analyses of some of
the rocks are given in Table 5.4, which indicates that the SiO_2 content
varies from 45% to about 56%; the Al_2O_3 and CaO contents are rather

Table 5.3. Chemical analyses of leucitites from New South Wales, Australia. Data were obtained from Cundari (1973)

	1	2	3	4	5
SiO_2	42.94	41.91	45.90	45.20	44.37
TiO_2	5.51	5.40	5.26	4.10	3.74
Al_2O_3	8.72	7.58	8.95	9.20	8.22
Fe_2O_3	4.98	7.07	5.09	4.79	4.87
FeO	6.31	4.24	5.56	6.39	5.70
MnO	0.15	0.15	0.13	0.17	0.17
MgO	12.17	13.31	9.71	11.80	14.43
CaO	9.30	9.09	7.86	8.98	9.01
Na_2O	0.74	0.92	1.21	2.49	1.94
K_2O	5.14	5.44	7.24	4.48	5.11
H_2O^+	2.60	2.36	1.32	0.40	0.88
P_2O_5	0.87	1.19	0.67	1.19	1.02
CO_2	–	–	–	–	–
σ	0.76	1.25	0.46	0.48	0.54
Total	100.19	99.91	99.36	99.67	100.00
(ppm)					
Cr	212	317	266	407	573
Ni	290	424	351	348	425
Co	36	39	41	40	43
Zr	657	740	700	520	570
Rb	157	293	299	86	112
Sr	2,100	2,190	830	1,160	1,300
Ba	2,950	6,800	1,350	1,340	1,420

σ: Σ Trace elements converted to per cent
The rocks are from the following localities: (1) Byrock (BQ–5), (2) El Capitan (CPT–4), (3) El Capitan (CPT–9), (4) Condoblin (CND–4), (5) Condoblin (CND–6)

high, whereas the TiO_2 contents are very low. Mineralogically these rocks are characterized by leucite, augite, K-feldspar, and calcic plagioclase in various proportions. Nepheline is usually absent in the rocks of Mt. Mouriah, whereas olivine and biotite are occasionally present.

In Celebes olivine leucitite, leucitite, leucite trachyte, and leucitite tuffs occur at various localities along the coast from Cape Mandar to Cape William and near the Bay of Mamudju. Leucite basanite is found in Oldeidu Kiki in the Matinang Mountains, southeast of Bowool in Northern Celebes. These lavas are associated with trachytes and trachyandesites. Potassium-rich rocks have also been reported in Pic de Maros Mountain, located between Maros and Tjambo, north of Makassar in Celebes. These rocks occur as intrusive bodies of shonkinites, consist-

Table 5.4. Analyses of leucite-bearing rocks from Java and Celebes, Indonesia

	1	2	3	4	5	6
SiO_2	50.18	48.66	46.54	54.97	46.04	46.60
TiO_2	0.76	0.81	1.11	–	2.20	0.95
Al_2O_3	17.82	17.69	15.95	22.21	12.40	16.73
Fe_2O_3	4.04	4.66	5.24	0.61	3.54	4.17
FeO	3.89	4.40	5.51	–	5.58	4.78
MnO	0.30	1.49	0.18	–	Trace	0.41
MgO	2.88	3.03	4.70	0.26	12.60	4.65
CaO	7.19	6.43	10.09	0.49	8.38	10.82
Na_2O	3.29	3.98	2.28	0.81	1.62	2.62
K_2O	6.65	6.10	4.44	19.98	4.87	5.47
H_2O^+	0.96	0.80	0.52	0.08	3.55	0.71
H_2O^-	0.55	0.58	0.59	0.56	–	0.45
P_2O_5	0.76	0.79	1.18	–	–	1.50
BaO	0.25	0.16	0.13	–	–	0.21
SrO	0.29	0.21	0.24	–	–	0.13
S	0.02	0.05	0.09	–	–	0.01
Cl	0.16	0.24	0.07	–	–	0.08
F	0.02	0.16	0.06	–	–	0.17
Total	100.01	100.24	98.93	99.97	100.78	100.46

1. Leucite tephrite from a locality near Ragau, Java
2. Leucite tephrite from Gillinan, Java
3. Leucitophyre from Gillinan, Java. Includes 0.01% ZrO_2
4. Leucite tephrite from Kali Gillinan, Java
5. Mica-bearing olivine leucitite from Oeloe Kajan, E. Borneo
6. Leucite tephrite from a locality near Ragau, Java
1-6 are analyses of rocks from Java and Borneo (Iddings and Morley 1915)

ing of augite, K-feldspar, biotite, and rare plagioclase. In some localities such as Gentungen, the shonkinites contain pseudoleucite and are very similar to the shonkinitic rocks of Montana, USA. These shonkinites grade into nepheline syenite, syenite porphyry, and trachyte.

5.6 Volcanic Fields of Highwood Mountains, Montana (USA)

The Highwood Mountains lie about 150 km southwest of Bearpaw Mountains (Figs. 5.1 and 5.7). The rock types of the latter area are also similar to those of the Highwood Mountains. Larsen et al. (1941) recognized four petrographic subprovinces in this area:

1. A series of quartz latites, high in K_2O and MgO and lower in Al_2O_3 than calc-alkalic lavas.

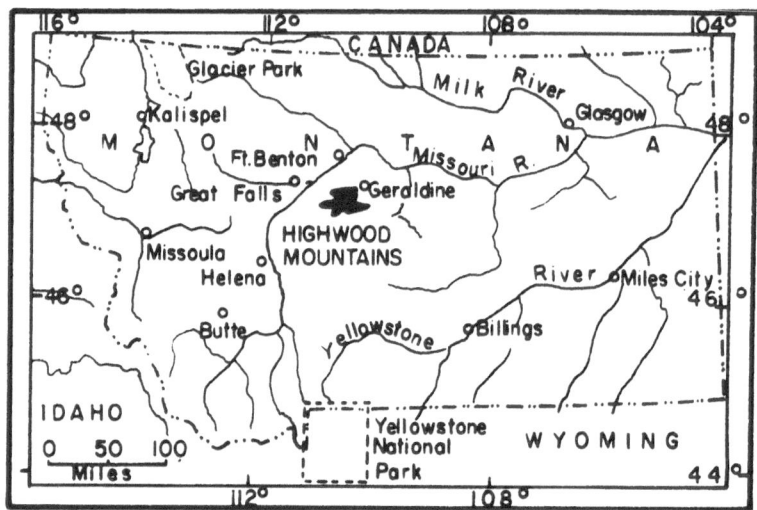

Fig. 5.7. Location of the leucite-bearing locality of the Highwood Mountains region. (After Larsen et al. 1941)

2. An intrusive stock of monzonite and syenite.
3. A small intrusion of ultramafic rock (monticellite peridotite and alnöite), rich in CaO and BaO.
4. Mafic potassic rocks and their derivatives.

The last subprovince is noted for abundant flows and dikes of phonolites and laccoliths and dikes of shonkinites. According to Larsen et al. igneous activity in the area was initiated by the eruption of quartz latites, which built up a volcanic pile of over 45 km in diameter on an irregular surface of Late Cretaceous sediments. This was followed by renewed volcanism of latite and/or analcite-bearing mafic phonolite, mafic analcite phonolite, mafic pseudoleucite phonolite, and mafic phonolites (without phenocrysts of leucite or analcite). Leucite-bearing phonolites in particular occur as thick flows in the northeastern and southern parts of the volcanic edifice.

The ferromagnesian minerals include augite, olivine, and mica. Feldspars are usually sanidine, containing appreciable amounts of barium. Analyses of some of the leucite phonolites from this area show that, like potassium-rich rocks of other areas, these rocks are enriched in Cl, P_2O_5, BaO, and SrO (Table 5.5).

Shonkin Sag laccolith (Nash and Wilkinson 1970, 1971), which occurs in this petrographic subprovince is essentially composed of pseudo-

Table 5.5. Analyses of potassic rocks from Highwood Mountains and Navajo-Hopi Province, USA

	1	2	3	4	5	6
SiO_2	47.98	46.62	47.50	52.80	53.60	54.11
TiO_2	0.54	0.76	1.85	1.58	1.39	1.71
Al_2O_3	13.34	12.48	9.62	11.14	11.18	11.20
Fe_2O_3	4.09	4.78	3.37	3.57	3.53	4.26
FeO	4.24	4.44	4.74	3.35	3.21	2.52
MnO	Trace	0.09	Trace	0.13	0.11	0.11
MgO	7.01	8.90	13.00	7.99	7.47	7.64
CaO	9.32	11.94	9.00	7.97	7.51	6.36
K_2O	5.00	4.42	3.28	6.71	7.14	6.05
Na_2O	3.51	1.97	1.96	2.10	1.98	2.50
H_2O^+	2.10	2.83	0.90	1.33	1.51	1.88
H_2O^-	–	–	3.90	0.37	0.16	0.97
BaO	0.50	0.60	–	–	–	–
SrO	0.04	0.04	–	–	–	–
P_2O_5	1.03	0.18	1.05	1.00	1.02	0.76
CO_2	1.24	–	–	–	–	–
Cl	0.21	–	–	–	–	–
F	–	–	–	–	–	–
Total	100.15	100.05	100.17	100.04	99.81	100.07

1, 2. Leucite phonolite, Highwood Mountains (Larsen et al. 1941; p. 1749)
3. Olivine leucitite, Navajo-Hopi Province (Williams 1936; p. 166)
4. Minette, The Beast, Arizona (neck) (K. Aoki personal communication)
5. Minette, Outlet Neck, Arizona (K. Aoki personal communication)
6. Biotite trachybasalt, Washington Pass, Arizona (K. Aoki personal communication)

leucite-bearing shonkinite. A lensoid sheet of syenite, also containing pseudoleucite, occurs at the top of the laccolith. Alkali feldspar and zeolite make up 70% of the total composition of this syenite, which also includes small to moderate amounts of augite and accessory olivine.

5.7 Potassic Rocks of Navajo-Hopi Province (USA)

Williams (1936) described potassium-rich volcanic rocks from Hopi Butte, Arizona and Navajo Mountain, Utah (Fig. 5.1). In Hopi Butte the alkaline rocks occur as pyroclastics and lava flows of limburgites, monchiquites, and nepheline-bearing trachytes. The eruption of lavas in this area was considered by him to have taken place sometime between Late Pliocene and Pleistocene.

Volcanism in the Washington Pass region of Navajo has been described more recently by Ehrenberg (1977). In this area eruption began with ejection of voluminous pyroclastic deposit including both tuff breccias, composed of basement fragments and agglomerates of minette clasts. In the waning stage of pyroclastic activity trachybasalt flows covered the crater floor, whereas vitreous to aphanitic minettes were emplaced in the crater center. The minettes consist of phlogopite, diopside, and olivine (Fo_{80}) phenocrysts in a matrix of biotite, clinopyroxene, and Ti-magnetite microphenocrysts with interstitial sanidine, analcite, and chlorite. In the final stage of volcanic activity there was again extrusion of a small volume of trachyte. The trachyte consists of biotite and diopside phenocrysts in a groundmass of sanidine laths and interstitial brown clay-like material.

Williams (1936) described the occurrence of olivine leucitite from Todilto Park of this petrographic province. Chemical analyses of some of the potassic rocks from this area as obtained by Williams, and Aoki (pers. comm. 1977) are listed in Table 5.5.

5.8 Leucite-Bearing Rocks of Leucite Hills, Wyoming (USA)

Leucite Hills is situated in the southwestern part of the state of Wyoming about 100 km north of the border of the states of Colorado and Wyoming. The outcrops of leucite-bearing rocks extend over a distance of 45 km along east-west trend from Black Rock Mesa to Pilot Mesa in a belt some 40 km wide. The area has been studied by Cross (1897), Kemp (1897), Kemp and Knight (1903), Yagi and Matsumoto (1966), and Carmichael (1967). According to these investigators, the eruption of potassic lavas occurred in early Tertiary times. The associated sedimentary rocks include sandstone, oolitic limestone, limestone, and shale. The potassic rocks occur as flows, intrusive sheets, dikes, and volcanic necks. Many of these flows with topographic expression of flat-topped hills (Mesa), are highly amygdoloidal. Kemp and Yagi and Matsumoto noted that micas in these lavas often have parallel arrangements, presenting a pseudo-schistose structure.

The most predominant rock types of the area are orendites and wyomingites. Of the twenty two localities of leucite-bearing rocks recorded by Kemp, the following areas were described by Carmichael to be the type localities of wyomingites: Boars Tusk, South Table Mountain, Zirkel Mesa, North Table Mountain, and Steam Boat Springs. Some of the important localities of orendites include South Table Mountain and Orenda Butte. According to Cross, xenoliths associated with the potassic rocks are not frequent and consist of both foreign and cognate blocks. Partially fused and friable accidental xenoliths of granites and arkose have been described by Carmichael. Yagi and Matsumoto recorded cognate xenoliths of leucitite in wyomingite and orendite from the Emmons Cone area. Chemical analyses of leucite-bearing rocks and some cognate xenoliths are given in Table 5.6.

Table 5.6. Analyses of leucite-bearing rocks from Leucite Hills, Wyoming, USA

	1	2	3	4	5	6	7
SiO_2	50.23	55.43	54.17	43.56	48.94	52.64	48.22
TiO_2	2.30	2.64	2.67	2.31	1.76	1.72	1.78
Al_2O_3	10.15	9.73	10.16	7.85	12.44	13.38	12.99
Fe_2O_3	3.65	2.12	3.34	5.57	4.28	5.19	2.81
FeO	1.21	1.48	0.65	0.85	3.71	1.63	5.83
MnO	0.09	0.08	0.06	0.15	0.10	0.09	0.10
MgO	7.48	6.11	6.62	11.03	5.84	4.40	5.83
CaO	6.12	2.69	4.19	11.89	4.77	3.16	6.16
Na_2O	1.29	0.94	1.21	0.74	2.17	2.22	1.93
K_2O	10.48	12.66	11.91	7.19	11.01	11.96	10.57
H_2O^+	2.34	2.07	1.01	2.89	1.43	1.87	1.36
H_2O^-	1.09	0.61	0.52	2.09	0.54	0.44	0.56
P_2O_5	1.81	1.52	1.59	1.50	0.47	0.44	0.50
Cr_2O_3	0.06	0.02	0.05	0.04	0.09	0.03	0.12
V_2O_3	–	–	–	–	0.06	0.01	–
BaO	0.61	0.64	0.59	0.66	0.81	0.37	–
SrO	0.32	0.27	0.18	0.40	0.38	0.22	–
ZrO_2	0.25	0.28	0.22	0.27	–	–	–
SO_3	0.35	0.46	0.16	0.52	0.44	0.09	0.64
Cl	–	–	0.06	–	0.05	0.03	0.05
F	–	–	0.36	–	0.71	0.54	0.69
Total	99.83	99.75	100.21	99.51	100.11	100.43	100.24

1. Wyomingite, Margin dike, Boars Tusk
2. Wyomingite, partly glassy lava, Steamboat Springs
3. Orendite, North Table Butte, Cross, 1897. It also includes 0.49% CO_2
4. Madupite, west side Pilot Butte, Rock Springs
5. Wyomingite, Emmons Cone. It also includes 0.11% ZnO
6. Orendite, Emmons Cone
7. Leucitite xenolith (1101 x) within a wyomingite, Emmons Cone

Analyses of 1 to 4 are from Carmichael (1967, p. 50) and those of 5 to 7 are from Yagi and Matsumoto (1966)

5.9 Leucite-Bearing Rocks of Italy

Widespread volcanic activity in the Roman Province of Italy has produced a variety of lavas, ignimbrites, and pyroclastic materials during
the Quaternary. Predominantly potassic lavas were erupted from a
number of centers within the Roman Comagmatic Region, which
covers an area of about 6000 km² extending from Lago di Bolsena in
North Lazio to Vesuvius and the Phlegrean Fields in the southeast
(Fig. 5.8). According to Marinelli and Mittempergher (1966) the basement rocks of this volcanic province are essentially made up of Mesozoic carbonates and flysch-type sediments, which are characterized
chiefly by the presence of marl and shale-bearing formations with sandstone and conglomerates. These rocks are distributed along lines of step
faulting and large-scale tectonic collapses. Washington (1906) divided
the area into various volcanic districts including: (a) Monti Vulsini,
(b) Vico, (c) Monti Sabatini, and (d) Colli Albani. In the Naples area
there is another set of volcanic centers: (e) Roccamonfina, (f) Phlegrean

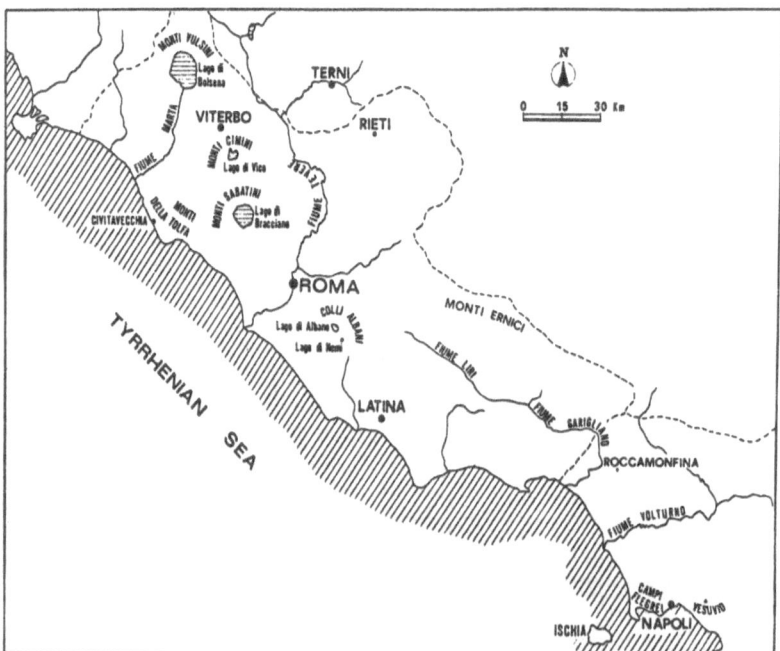

Fig. 5.8. Location of the volcanic fields of Italy, noted for leucite-bearing rocks.
(After Turi and Taylor 1976)

Fields, and (g) Ischia. Only the areas noted for leucite-bearing rocks are described in the following:

5.9.1 Monti Vulsini

This volcanic district is the northernmost locality of the Roman Co-magmatic Volcanic Province (Fig. 5.9). It covers an area of about 2300 km² around the Lago di Bolsena Caldera (Turi and Taylor, 1976). The area has been studied by Washington (1906), Mattias (1965), Schneider (1965), Nappi (1969), and Trigila (1969). According to Evernden and Curtis (1965), K/Ar dating of the rocks gave ages varying from 275,000 to 431,000 years.

The rock types consist of leucite basanites, leucite tephrites, trachytic ignimbrites, latites, and trachytes. Chemical analysis of a leucite trachyte is given in Table 5.9.

Locardi and Sircana (1967) described the activity of a later volcano from the Monti Vulsini district, which lies west of Lago di Bolsena. Three main stages of eruption have been characterized by the following rock types:

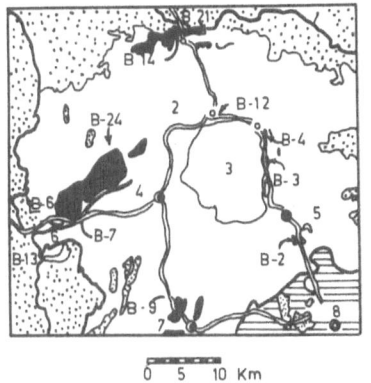

B-2 Zepponami Via Cassia km 947

B-3 Via Cassia km 109.5

B-4 Pietre Lanciate Via Cassia km 111.4

B-6 Selva del Lamone Valentano-Manciano Road km 13.0

B-7 Bridge on Fosso Olpeta Manciano Road km 11.2

B-9 Tuscania-Valentano Road km 50

B-12 San Lorenzo Nuovo Via Cassia km 123.7

B-13 Monte Calvo

B-14 Acquapendente Via Cassia km 133.0

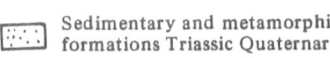

0 5 10 Km

Sedimentary and metamorphic formations Triassic Quaternary

 Ciminian and Vican volcanics

Vulsinian volcanics

1. F. Paglia, 2. San Lorenzo Nuovo, 3. Lago di Balsena, 4. Valentano, 5. Montefiascone,

6. F. Olpeta, 7. Tuscania, 8. Viterbo

Fig. 5.9. Location of Monti Vulsini and adjoining areas. (After Turi and Taylor 1976)

1. Alternation of tuffs and leucitite lavas.
2. Trachytic ignimbrites, followed by latitic and leucitic trachyte flows.
3. Scoria and ash flows, followed by latitic and leucititic lavas.

Discendenti et al. (1970) determined the ages of some pumices, lavas, and sanidine-bearing inclusions of later volcano by K/Ar and ^{230}Th methods. The ages of the rocks are found to range from 55,000 to 19,000 years.

5.9.2 Vico Volcanic Complex

This volcanic complex, situated in the northern part of the Roman Province, was studied by Mattias and Ventriglia (1970; Fig. 5.10) and more recently by Cundari and Mattias (1974) and Cundari (1975). According to Cundari and Mattias it is a stratovolcano, forming a lobate lake-filled caldera with a nested cone. The volcanic complex is made up of pyroclastic materials with six distinct formations, which were produced during the major episode of central vent or vent formation.

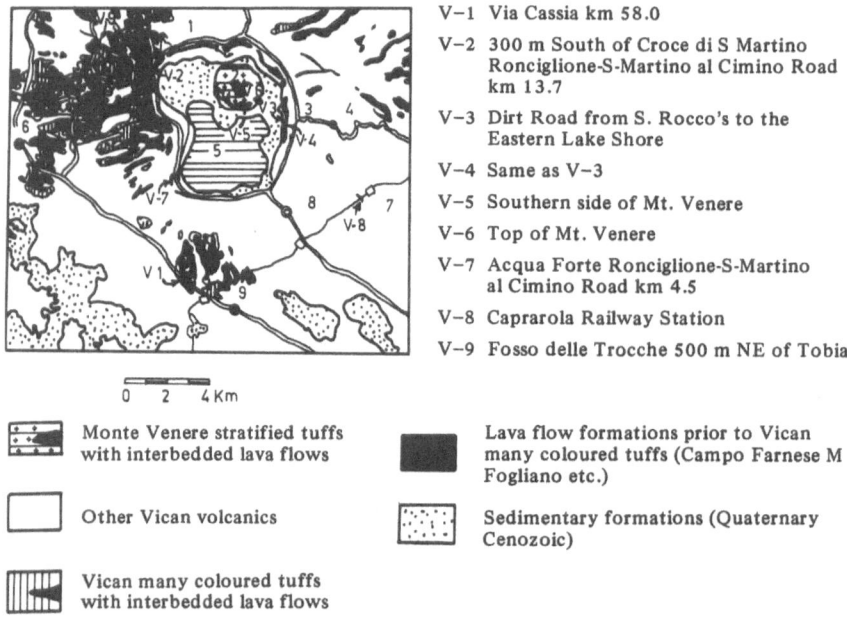

V-1 Via Cassia km 58.0

V-2 300 m South of Croce di S Martino
     Ronciglione-S-Martino al Cimino Road
     km 13.7

V-3 Dirt Road from S. Rocco's to the
     Eastern Lake Shore

V-4 Same as V-3

V-5 Southern side of Mt. Venere

V-6 Top of Mt. Venere

V-7 Acqua Forte Ronciglione-S-Martino
     al Cimino Road km 4.5

V-8 Caprarola Railway Station

V-9 Fosso delle Trocche 500 m NE of Tobia

0   2   4 Km

Monte Venere stratified tuffs
with interbedded lava flows

Other Vican volcanics

Vican many coloured tuffs
with interbedded lava flows

Lava flow formations prior to Vican
many coloured tuffs (Campo Farnese M
Fogliano etc.)

Sedimentary formations (Quaternary
Cenozoic)

1. San Martino al Cimino, 2. M. Venere, 3. S. Racco, 4. Caprarola, 5. Lago di Vico,
6. Vetralla, 7. Stazione di Caprarola, 8. Ranciglione, 9. Capranica

Fig. 5.10. Location of Vico volcanic complex. (After Turi and Taylor 1976)

Most lavas are intercalated with well-stratified and weakly cemented pyroclastic deposits, formed by repeated layers of ash, pumice, and lapilli, reaching a maximum thickness of 30 m. Plotting the chemical analyses of the rocks from this area in an A-P-F diagram, Cundari and Mattias found that most of the compositions lie within the tephritic leucite phonolite field (TLP). The distribution grades from tephritic leucitite (TL) through phonolitic leucite tephrite (PLT) to alkali trachyte (AT, Fig. 5.11). In Table 5.7 analyses of different leucite-bearing rocks from the Vico area are given. On the basis of their work Cundari and Mattias concluded that the composition of the possible parent magma would plot in the TL-PLT-TLP field (Fig. 5.11).

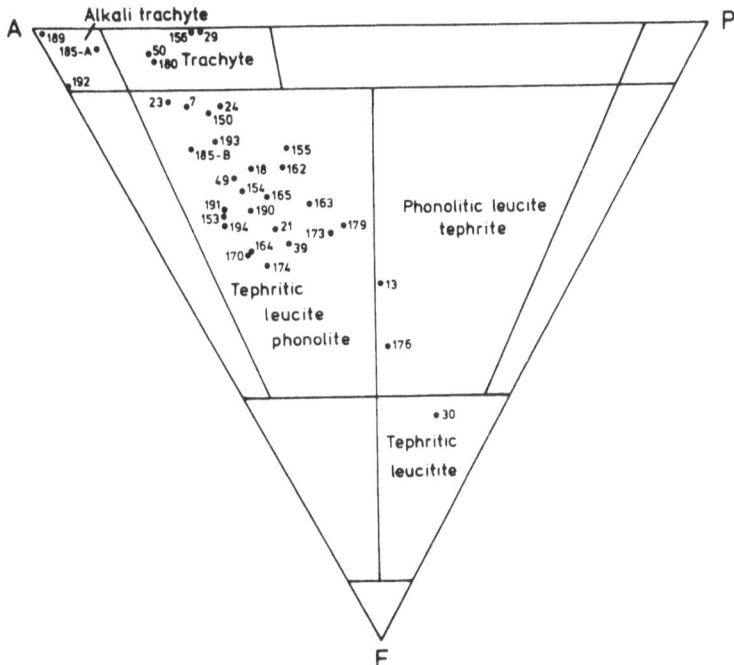

Fig. 5.11. Plot of the modal composition of leucite-bearing rocks in an *A* (alkali feldspar) – *P* (plagioclase) – *F* (feldspathoid) diagram. (After Cundari and Mattias 1974)

5.9.3 Monti Sabatini

This area is a multicentered volcanic district (Turi and Taylor 1976), which includes Lago di Martignano and Lago di Bracciano (Fig. 5.12).

Table 5.7. Analyses of leucite-bearing rocks from the Vico volcanic area, Italy

	1	2	3	4	5	6	7
SiO_2	55.0	53.1	56.2	52.0	51.8	51.4	49.0
TiO_2	0.59	0.70	0.58	0.67	0.80	0.71	0.80
Al_2O_3	16.8	17.9	19.5	16.5	17.1	19.2	18.3
Fe_2O_3	2.6	3.6	1.8	2.3	2.3	2.8	3.6
FeO	3.0	2.8	1.7	3.9	4.8	4.3	3.4
MnO	0.13	0.13	0.10	0.15	0.15	0.17	0.15
MgO	3.6	3.3	1.7	5.3	3.7	2.9	3.5
CaO	6.2	6.1	3.7	7.8	7.2	6.4	8.3
Na_2O	1.5	2.4	2.5	2.2	1.8	2.4	1.3
K_2O	8.4	6.6	9.8	6.3	8.4	8.1	8.8
H_2O	1.48	2.10	1.58	1.26	1.10	0.88	1.56
P_2O_5	0.45	0.48	0.35	0.73	0.63	0.60	0.60
CO_2	0.12	0.02	0.10	0.04	0.06	0.10	0.22
σ	0.51	0.39	0.55	0.40	0.51	0.44	0.86
Total	100.38	99.62	100.16	99.55	100.35	100.40	100.39
(ppm)							
Zr	396	469	520	365	475	334	551
Ba	2,047	1,165	2,056	1,493	1,801	1,437	3,659
Sr	1,401	1,205	1,440	1,096	1,525	1,407	2,590
Rb	539	584	711	416	602	553	627
K/Rb	130	94	114	125	116	121	116
Rb/Sr	0.38	0.48	0.49	0.38	0.39	0.39	0.24

σ = Trace elements converted to percent
1 (LTR 189) and 2 (LTR 29) are leucite-bearing trachytes. 3 (TLP 185–B), 4 (TLP 154), and 5 (TLP 21) are tephritic phonolites. 6 (PLT 13) and 7 (PLT 176) are phonolitic tephrities. All analyses were taken from Cundari and Mattias (1974)

The latter is located on a large volcanic tectonic depression of about 9 km in diameter. The volcanic rocks of this area vary from trachytes to leucitites. A considerable portion of this area is covered by the products of the Vico volcano. The periods of volcanic activity in this area are as follows:

1. The first period is represented by tephritic ignimbrites and lavas older than 820,000 years, preceding the activity of Vico.
2. The second period of volcanic activity is contemporaneous with the Vico eruption. According to Evernden and Curtis (1965) K/Ar methods gave ages of these rocks between 438,000 and 417,000 years.
3. The third period was characterized by strongly silica-undersaturated rocks such as tephrites and leucitites, which according to Turi and

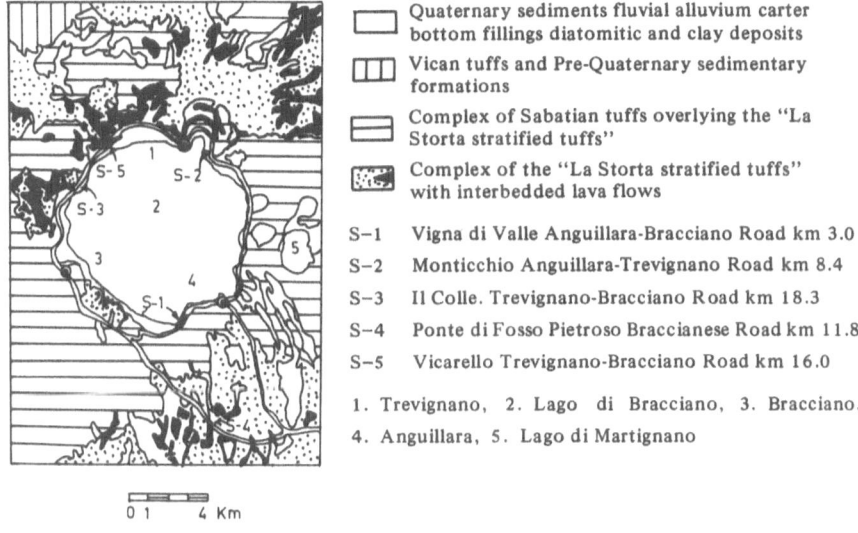

Quaternary sediments fluvial alluvium carter bottom fillings diatomitic and clay deposits

Vican tuffs and Pre-Quaternary sedimentary formations

Complex of Sabatian tuffs overlying the "La Storta stratified tuffs"

Complex of the "La Storta stratified tuffs" with interbedded lava flows

S-1   Vigna di Valle Anguillara-Bracciano Road km 3.0

S-2   Monticchio Anguillara-Trevignano Road km 8.4

S-3   Il Colle. Trevignano-Bracciano Road km 18.3

S-4   Ponte di Fosso Pietroso Braccianese Road km 11.8

S-5   Vicarello Trevignano-Bracciano Road km 16.0

1. Trevignano, 2. Lago di Bracciano, 3. Bracciano, 4. Anguillara, 5. Lago di Martignano

0 1   4 Km

Fig. 5.12. Location of Monti Sabatini volcanic field. (After Turi and Taylor 1976)

Taylor, are younger than 300,000 years in age. Widespread lava flows of the summit caldera of Monti Sabatini also erupted during this period.

5.9.4 Colli Albani (Alban Hills)

This volcanic district (Fig. 5.13) was studied by Sabatini (1900), Washington (1906), Fornaseri et al. (1963), and Ambrosetti et al. (1972). Lava flows from this area consist of nepheline leucitite, nepheline-melilite leucitite, olivine leucitite, tephritic leucitite, and leucitite, and their ages vary from 43,000 to 700,000 years (Ambrosetti et al., 1972). The lavas often include xenoliths and ejecta of biotite leucitite, magnetite leucitite, and biotite xenocryst.

5.9.5 Phlegrean Fields

The Phlegrean Fields are located around Naples and cover an area of 15-20 km in diameter, bounded by the Sebeto River to the east and by the Volturno plain to the north. Many centers are found in this area, the activity of which possibly started in the Pliocene. The first eruptions took place through faults extending in the east-northeast and west-southwest direction (Tyrrhenian fault). Following this activity

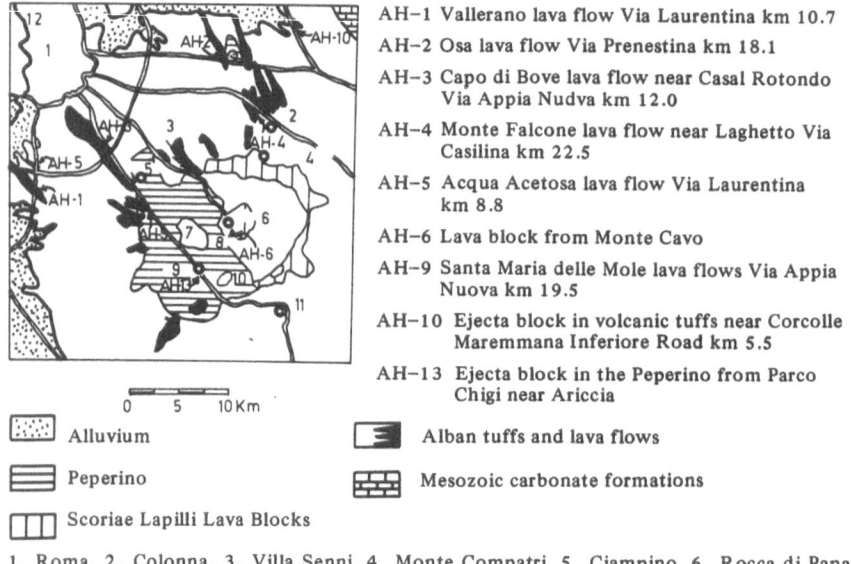

AH-1 Vallerano lava flow Via Laurentina km 10.7

AH-2 Osa lava flow Via Prenestina km 18.1

AH-3 Capo di Bove lava flow near Casal Rotondo Via Appia Nudva km 12.0

AH-4 Monte Falcone lava flow near Laghetto Via Casilina km 22.5

AH-5 Acqua Acetosa lava flow Via Laurentina km 8.8

AH-6 Lava block from Monte Cavo

AH-9 Santa Maria delle Mole lava flows Via Appia Nuova km 19.5

AH-10 Ejecta block in volcanic tuffs near Corcolle Maremmana Inferiore Road km 5.5

AH-13 Ejecta block in the Peperino from Parco Chigi near Ariccia

0   5   10 Km

[:::::] Alluvium   [≡] Alban tuffs and lava flows

[≡] Peperino   [⊞] Mesozoic carbonate formations

[⊞] Scoriae Lapilli Lava Blocks

1. Roma, 2. Colonna, 3. Villa Senni, 4. Monte Compatri, 5. Ciampino, 6. Rocca di Papa,
7. L. di Albano, 8. M. Cavo, 9. Ariccia, 10. L. di Nemi, 11. Velletri, 12. F. Tevere

Fig. 5.13. Location of Colli Albani Volcanic field. (After Turi and Taylor 1976)

there occurred several cycles of volcanism. The main historic eruption of the Phlegrean Fields took place from Mt. Nuvo, where thermal activity can still be observed in the form of fumaroles at Solfatara. Quite recently volcanic activity, accompanied by upheaval of the ground, occurred in 1970. During the Pliocene the igneous activity began with the eruption of trachybasalt, and different magma types were produced through differentiation within the conduit (Imbo 1965), erupting trachytes, sodalite-bearing phonolites and trachyandesites. According to Imbo, leucite-bearing tephrites are also known from the Phlegrean Fields.

5.9.6 Somma-Vesuvius

The Somma-Vesuvius volcanic complex is located to the east of Naples in the Campanian Plain. The volcanic complex is bounded to the east by the Apennine range for a distance of about 10 to 40 km. In the west it is partially limited by the Bay of Naples and to the north by the Phlegrean Fields. During the eruptions of the Pleistocene, the first lava flow covered the Phlegrean Fields volcanics. The activity was followed by the alternate eruptions in Somma-Vesuvius and Phlegrean Fields volcanic complexes. According to Imbo the volcanic eruptions of Somma-Vesuvius can be divided into four stages: (1) Pre-Somma, (2) Older Somma, (3) Younger Somma, and (4) Vesuvius.

The first historic eruption took place in 79 A.D., after which there have been numerous eruptions throughout the centuries, each time followed by a period of quiescence. The last three eruptions took place between 1914 and 1948. According to Imbo, during these eruptions, there were forty five events following one after another until the concluding paroxysm on March 18, 1944. Each event was characterized by the following phases: fracture formation on the little cones, explosive activity involving ejection of incandescent products and rebuilding of the cones. The change in the morphology of the crater of Vesuvius is shown in Fig. 5.14.

The main components of the lavas of the Somma-Vesuvius region are augite, plagioclase, and sanidine. Olivine, biotite, sometimes apatite, and opaque minerals are less abundant. The glass content of the lavas of Somma-Vesuvius region is variable, sometimes reaching as high as 77% in some pumices ejected over Pompei during the Plinian eruption of 79 A.D. Augite is the only pyroxene in the great majority of Somma-Vesuvius lavas, whereas diopside is common in the Phlegrean trachytes. In the trachytes of Cuma (near Grotta della Sibilla) aegirinaugite substitutes for augite. Rahman (1975) analyzed clinopyroxenes from the Somma-Vesuvius region (Table 3.6.).

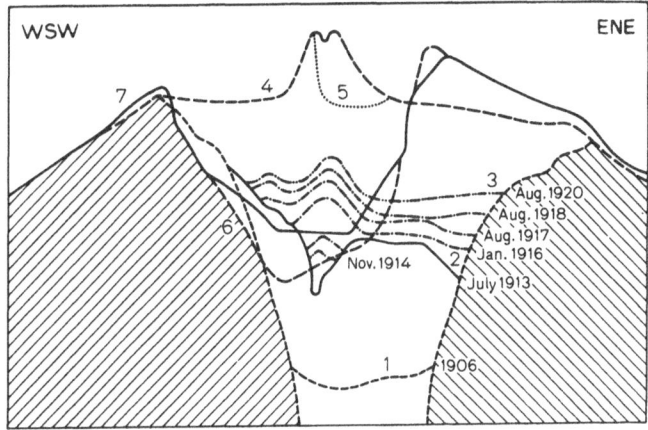

Fig. 5.14. Change in the morphology of the crater of Vesuvius. (After Imbo 1965)

Table 5.8. Analyses of potassic rocks from the Phlegrean Fields and the Somma-Vesuvius region of Italy

	1	2	3	4	5	6	7
SiO_2	58.9	48.7	52.1	48.3	51.7	51.4	50.5
TiO_2	0.33	0.78	0.37	0.63	0.67	0.65	0.8
Al_2O_3	19.6	19.3	19.9	19.8	19.0	18.6	17.8
Fe_2O_3	1.1	3.70	2.20	2.0	2.6	2.8	3.0
FeO	2.3	4.3	1.6	5.2	4.0	3.9	4.6
MnO	0.10	0.12	0.13	0.12	0.10	0.11	0.12
MgO	1.3	3.4	1.2	2.6	4.4	4.3	5.1
CaO	2.9	8.3	3.2	8.1	7.8	8.5	10.6
Na_2O	3.9	3.3	4.5	3.1	2.5	3.6	2.5
K_2O	7.1	7.0	8.4	7.9	5.5	3.8	4.2
H_2O^+	1.34	0.53	3.0	0.3	0.23	1.29	0.7
H_2O^-	0.18	0.15	0.8	0.06	0.21	0.25	0.3
P_2O_5	0.15	0.64	0.31	0.64	0.6	0.6	0.68
BaO + SrO	0.1	0.37	0.16	0.41	0.28	0.27	0.24
Cl-O-Cl	0.11	0.42	0.53	0.31	–	–	0.02
F-O-F	0.06	0.21	0.26	0.2	0.03	0.15	0.15
Total	99.47	101.22	98.66	99.67	99.62	100.22	101.31
(ppm)							
F	1,000	3,600	4,500	3,500	500	2,600	2,500
Cl	1,400	5,400	6,800	4,000	50	50	350
Mn	800	940	1,000	920	780	820	910
Cu	40	150	55	110	120	85	90
Zn	70	90	90	75	75	70	75
Rb	330	230	340	340	260	340	230
Sr	490	900	780	1,000	630	720	660
Zr	460	320	250	320	240	250	240
Nb	30	10	10	20	30	45	10
Ba	310	2,300	21,500	2,600	1,800	1,700	1,400

1. Trachyte (CF 45), south wall of Solfatara
2. Trachytic leucitite (SV 25), in a road side from Torre Annunziata to Boscotrecase
3. Vitreous pumice over Pompei, 79 A.D. eruption (SV 30)
4. Vesuvite (SV 40), Valle del Gigante
5. Phonolitic tephrite (SV 10), Canolone of Pollena
6. Phonolitic tephrite (SV 13), Canolone of Pollena
7. Phonolitic tephrite (SV 17), Castello di Cisterna

7 is from the Phlegrean Fields, whereas other rocks are from the Somma-Vesuvius area. All analyses were obtained from Savelli (1967)

The average modal compositions of the Somma-Vesuvius lavas are as follows: sanidine (24% So; 2% Ve); leucite (20% So; 38% Ve); plagioclase (26% So; 31% Ve). So represents rocks from Somma, whereas Ve represents Vesuvian rocks.

Chemical analyses of major and trace elements of leucite-bearing rocks from the Somma-Vesuvius region, as obtained by Savelli (1967) are given in Table 5.8. In his plot of MgO/SiO_2 contents in vesuvites in contrast to tephrites from Somma, the MgO content is lower in the former rocks than in the latter. The mean ratio of FeO/Fe_2O_3 is 1.37 in the Somma lavas but this ratio is 1.70 in vesuvites. The chemical compositions of the Somma-Vesuvius lavas are plotted in the A–P–F diagram by Savelli as shown in Fig. 2.2. Compositions of Somma-Vesuvius and Phlegrean Fields rocks are plotted with respect to nepheline, kalsilite, and SiO_2 in Fig. 5.15.

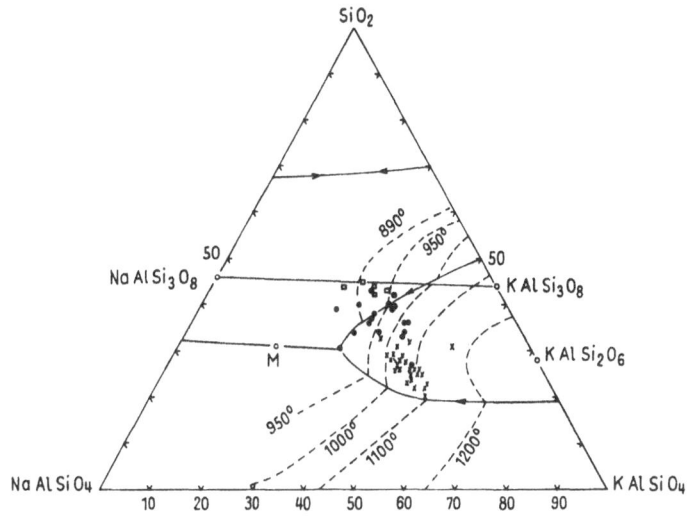

Fig. 5.15. Plot of the chemical composition of leucite-bearing rocks of Somma-Vesuvius and Phlegrean Fields in the system nepheline-kalsilite-SiO_2. Equilibrium diagram of the system $NaAlSiO_4$–$KAlSiO_4$–SiO_2–H_2O at P_{H_2O} = 1000 kg/cm^2 as determined by Hamilton and MacKenzie (1965). Normal compositions in wt. %. ● Somma lavas; x Vesuvian lavas; □ Phlegrean trachytes. M temperature minimum. (After Savelli 1967)

5.9.7 Roccamonfina

This is one of the Plio-Pleistocene volcanic centers situated to the west of the Appenines. The rocks of this locality fall distinctly into two series (Appleton 1972):

1. A highly potassic series, consisting of nepheline and leucite normative lavas.
2. A potassium-poor series, consisting of nepheline or quartz normative basalts, trachybasalts, and biotite augite latites.

Series 1 is richer in K, P, Ba, Ce, Rb, Sr, Th, and Zr than series 2. According to Appleton the highly potassic lavas of Roccamonfina can be further subdivided into the following groups: (a) clinopyroxene microphyric lavas, (b) leucite macrophyric lavas, (c) aphyric or microphyric lavas, (d) biotite clinopyroxene-plagioclase (An_{72-89})-magnetite phyric lavas, and (e) lavas containing sanidine as the main phenocryst phase. Group (c) lavas contain leucite, clinopyroxene, and plagioclase (An_{66-75}) as phenocrysts. Group (b) lavas contain phenocrysts of clinopyroxene, plagioclase, and magnetite with leucite ranging up to 40% by volume. Sanidine forms only in those lavas which contain more than 55% SiO_2.

Some of the leucite macrophyric lavas contain glomeroaggregates of clinopyroxene, plagioclase, magnetite with or without apatite, sometimes surrounded by leucite phenocrysts indicating an early period of crystallization of the glomeroaggregates prior to that of leucite.

Appleton concluded that an intermediate or high pressure process produced the parent magma with specific levels of enrichments in potassium and associated elements; and fractionation of biotite gabbro (which is found as nodules) from such a liquid at low pressure would produce a chemical variation towards salic derivants. Chemical compositions of some leucite-bearing rocks from this area are given in Table 5.9.

5.9.8 Volcanic Activity in the Eolian Arc Region

This Eolian arc includes seven islands: Lipari, Salina, Stromboli, Vulcano, Panarea, Filicudi, and Alicudi, which form an archipelago in the southern Tyrrhenian Sea, near the northern continental slope of Sicily. Barberi et al. (1974) described the volcanism of the Eolian Arc region in relation to an inclined seismic zone, developed along the boundary between the African and European plates. The volcanism in this region has been described in terms of plate tectonics by Ninkovich and Hays (1972). According to Barberi et al., the volcanism in this area is characterized by a marked evolution in a restricted period of 1 million years,

Table 5.9. Analyses of leucite-bearing rocks from Roccamonfina and Vulsinian district, Italy

	1	2	3	4	5	6
SiO_2	46.4	49.9	53.5	56.3	58.1	55.85
TiO_2	0.94	0.69	0.58	0.46	0.27	–
Al_2O_3	15.7	19.6	20.0	20.2	21.5	18.34
Fe_2O_3	5.3	2.9	2.8	2.9	1.6	3.77
FeO	3.4	3.9	2.5	0.8	0.6	1.88
MnO	0.15	0.16	0.15	0.13	0.15	–
MgO	6.0	2.5	1.6	0.9	0.2	1.73
CaO	11.6	6.9	5.2	3.6	2.0	3.84
Na_2O	1.6	2.4	3.4	3.2	4.7	3.39
K_2O	6.6	9.1	8.6	9.1	9.4	8.77
H_2O	0.94	0.85	0.93	1.66	1.07	1.14
P_2O_5	0.58	0.39	0.24	0.13	0.04	–
BaO	0.17	0.17	0.15	0.13	0.01	–
SrO	0.19	0.24	0.20	0.20	0.03	–
Total	99.57	99.70	99.85	99.71	99.67	98.71

1. Leucitite (average of 4 analyses)
2. Tephritic leucitite (average of 40 analyses)
3. Phonolitic leucite tephrite (average of 14 analyses)
4. Tephritic leucite phonolite (average of 8 analyses)
5. Leucite phonolite (average of 2 analyses)
6. Leucite trachyte, Bugnorea, Vulsinian district, Italy

1-5 are analyses of rocks from Rocca Monfina, Italy (Appleton 1972; p. 429).
6 is analysis from Washington (1906)

a progressive transition from typical calc-alkalic to shoshonitic rocks. The first volcanic stage was characterized by calc-alkalic series, ranging from high alumina basalt to dacite. At the beginning of the second stage, potassic andesites were emitted at Lipari, whereas in Salina the last stage of calc-alkalic volcanism took place. The most recent volcanism is characterized by further increase of the potassium content, indicating their shoshonitic associations, composed of shoshonitic basalts, grading to latites and trachytes. Barberi et al. considered that rhyolites from volcanello represent the final stage of this evolution. In Volcanello, leucite tephrites and related rock types were produced by fractionation of a shoshonitic magma under a P_{O_2} of 10^{-7} bar and the evolution of Eolian Volcanism might have been related to rapid deepening of the Benioff zone. The occurrence of shoshonitic rocks and the continental nature of the crust on both sides of the plate boundaries may suggest that the Eolian arc is in a senile stage of evolution. From their Sr iso-

topic data, Barberi et al. considered a mantle source of Eolian volcanism.

5.10 Volcanic Provinces of Spain

Leucite-bearing rocks in Spain occur in three volcanic provinces
(Fig. 5.16). The first province is in the southeast of Spain, where several
plugs, flows, and minor dikes of lamproitic rocks occur. The volcanic
activity in this region took place between middle Pliocene to post
Pliocene (A. Hernandez-Pacheo, personal communication; December
1977). The second province is situated in central Spain (Fig. 5.16),
outcrops of which are remnants of very eroded lava flows and volcanic
cones. The rock types range from basanites to trachybasalts, olivine
nephelinites, melilitites, and leucitites. The third volcanic province is
located near the Pyrenees in the Gerona Province, where some leucitic
rocks are reported. Ages of the rocks of this area range from Pliocene (?)
to Quaternary.

Detailed petrological and geochemical studies of the Mauricia and
Almaria regions (Fig. 5.16) have been made by Borley (1967) and
Caraballo (1975), who found verites and jumillites in widely scattered
localities. The verites form series of hills near Mauricia, and jumillites
occur as small hilly exposures near Hellin and Jumilla in Tertiary and
Quaternary sediments. The verites are found as massive black obsidian
or breccia and contain amygdules filled with calcite and quartz. Jumillites form Miocene lavas and are characterized by high K_2O/Na_2O ($>$ 2),
high MgO/CaO, higher concentration of Cr, Ti, Zr, and moderate
amounts of Ba (Table 5.10). Variation diagrams of the rocks of the
Leucite Hills, West Kimberley, and Spain are shown in Fig. 5.17. The
K_2O and Al_2O_3 contents of the rocks of the Leucite Hills and West
Kimberley are in general higher than those of the Spanish rocks.

5.11 Leucite-Bearing Lamprophyres of Eastern India

Leucite-bearing rocks are reported to occur as lamprophyre sills and
dikes in the Jharia and Bokaro coal fields of Bihar (Chatterjee 1974),
and in the Raniganj coal fields of West Bengal (Banerjee 1953). The
coal-bearing sedimentary formations of these localities were deposited
during the upper Paleozoic. However, lamprophyre intrusions probably

Fig. 5.16. Leucite-bearing petrographic provinces of Spain. *1* The town Murcia is located at a crowfly distance of about 230 km southeast of Madrid. *2* The town Cludad Real is situated at a crowfly distance of 160 km south of Madrid. *3* The town Gerona is located near the Gulf of Lions at a crowfly distance of about 550 km northeast of Madrid

Table 5.10. Analyses of potassic rocks from Spain

	1	2	3	4	5	6	7
SiO_2	53.96	53.24	55.79	51.40	45.53	45.64	47.07
TiO_2	1.30	1.24	1.33	1.12	1.57	1.51	1.32
Al_2O_3	11.3	10.8	11.4	10.6	8.5	8.2	7.2
Fe_2O_3	0.73	1.00	1.21	1.39	2.93	3.44	3.03
FeO	3.52	3.45	3.19	3.21	4.31	3.83	3.19
MnO	0.07	0.06	0.04	0.07	0.11	0.12	0.10
MgO	8.61	8.00	7.11	6.68	14.86	14.65	16.88
CaO	3.67	5.20	3.94	6.33	9.06	8.95	7.90
Na_2O	3.1	3.0	3.2	3.3	1.5	1.6	1.4
K_2O	3.5	3.4	3.8	3.4	3.6	3.8	5.0
H_2O^+	4.8	4.5	4.8	4.8	4.1	3.6	2.9
H_2O^-	0.47	0.46	0.50	0.66	1.13	0.66	0.52
P_2O_5	0.55	0.54	0.47	0.71	1.82	2.04	1.78
BaO	-	-	-	-	0.41	0.41	0.48
SrO	0.06	0.06	0.06	0.05	0.22	0.25	0.19
ZrO_2	0.08	0.09	0.10	0.09	0.12	0.10	0.09
Cr_2O_3	-	-	-	-	-	0.04	0.10
CO_2	4.4	5.3	3.6	6.7	0.6	0.8	1.2
Total	100.12	100.34	100.54	100.51	100.37	99.64	100.35

1-4. Analyses of verites from southern Spain
5-7. Analyses of jumillites from the same locality
All analyses were taken from Borley (1967; p. 373)

took place at a much later geologic period. In Sudamdih and Bhowra near Jharia, the lamprophyres consist of leucite, K-feldspar, phlogopite, and olivine (Chatterjee 1974). In the Raniganj and Bokaro coal fields the dikes have variable mineralogy corresponding to minette, kersantite, olivine-amphibole leucitite, biotite-olivine leucitite, and syenite.

5.12 Leucite-Bearing Rocks of the Laacher See and Other Areas of West Germany

The alkalic volcanic rocks of Germany can be divided into twelve volcanic districts (Wimmenauer 1974), of which the rocks of the Laacher See area are most alkalic. The area is located on the west side of the Rhine river (Fig. 5.18) about 40 km west of Koblenz. The whole area, including many localities in and around the Laacher See, is known as the East Eifel region. Ahrens (1961) considered that the volcanism in the East Eifel region was related to the development of the Neuwieder

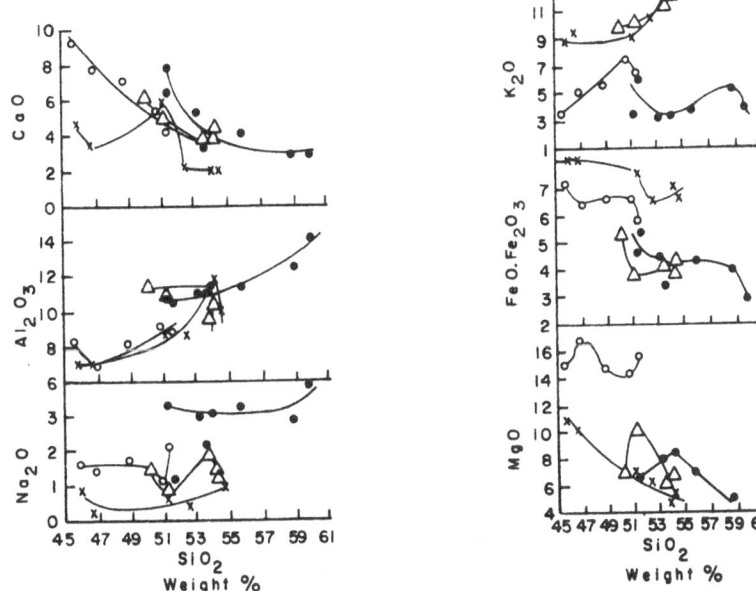

Fig. 5.17. Oxide variation diagrams for jumillites, verites, and related rocks. (After Borley 1967)

Basin. Cloos (1939) believed that the magma was erupted through a deep fissure system corresponding to the present Rhine Valley. Older volcanism took place about 420,000 years ago (Waal stage), whereas eruption of some of the younger lavas occurred about 11,000 years ago (Allerod Stage). The volcanic centers of pumiceous tuffs originally covered an area approximately 120 km long and 40 km wide. Highly potassic rocks of the East Eifel region occur as massive or vesicular lava flows, associated with tuffs and scoria and lapilli-filled layers, often containing bombs with chilled margins. Mineralogical compositions of the rocks of different localities of the East Eifel region are summarized in Table 5.11. The ages of the potassic rocks from this region are given in Table 5.12 (Frechen 1971). To the west of the East Eifel region lies the West Eifel volcanic district. In the latter area leucitic rocks occur at Steinraush (near Hill Shein), Erntsberg (near Hinterweilen), Hardt (near Mahren), Dockweil, Dockweiler, Kyllerkopf (lower lava), Gossberg (near Waldsdorf), Mosenberg, Döhmberg, and Alter Vos (near Berlinger). Chemical analyses of the rocks from the East Eifel are given in Table 5.13.

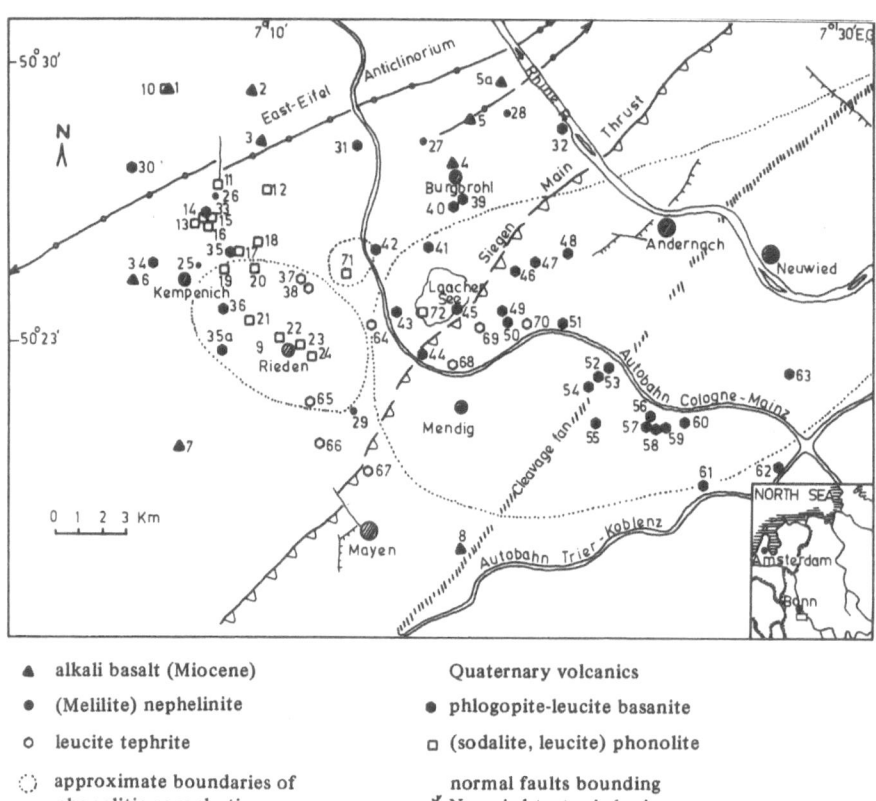

Fig. 5.18. Different localities of leucite-bearing rocks from the East Eifel. (After Duda and Schminke 1978)

Table 5.11. Mineralogical assemblages and rock types of various leucite-bearing localities of the East Eifel region

Rock name	Assemblage (groundmass minerals shown in brackets)	Locality
Basanite	Clinopyroxene, olivine, mica, and rare amphibole (plagioclase, leucite, nepheline, magnetite, apatite, and rare sphene)	Veitskopf; Nickenicher Hummerich; Heidekopf; Nickenicher Sattel; Sattelberg; Kunkskopf; Rothenberg; Alteburg; Roterberg

Table 5.11 (continued)

Rock name	Assemblage (groundmass minerals shown in brackets)	Locality
Tephrite	Clinopyroxene, mica, rare amphibole, rare olivine, and plagioclase (leucite, nepheline, magnetite, sphene, and apatite)	Epfelsberg; Rothenberg; Krufter Ofen
Nephelinite	Nepheline, augite, and rare olivine (mica, magnetite, sphene, and apatite; hauyne, melilite, and perovskite are also noted in rocks of some localities)	Herchenberg; Hannebacher Lei; Leilenkopf

Table 5.12. Ages of potassic rocks of the Laacher See area and their stratigraphic sequence. (After Frechen 1971)

Locality	Age (years B.C.)	Method
Tuff from Karlich	12,000–25,000	^{14}C
	25,000	
Herchenberg (young tuff)		
Kunkskopf		
Alteberg		
Mendiger lava		
Wingertsberg		
Thelenberg		
Krufter Ofen		
	32,000	^{14}C
Nickenicher group		
Nastberg		
Korretsberg		
Plaidter Hummerich		
Tonchesberg		
Wannen group I		
Karmalenberg group I		
	52,000	^{14}C
Herchenberg old tuff		
	59,000	^{14}C
Wannen group II		
Karmalenberg group II	62,000	^{14}C
Hohe Buch near Fornich	–	^{14}C
Bausenberg	140,000, 150,000	K–Ar
Veitskopf	–	
Lummerfeld	–	
Leilenkopf (young tuff)	220,000	K–Ar
Hochstein (young tuff)	–	K–Ar

Table 5.12 (continued)

Locality	Age (years B.C.)	Method
Hochsimmer	300,000	
Perlerkopf	320,000	
Engelnerkopf	320,000	K-Ar
Sulzbush	340,000	K-Ar
Hochstein (old tuff)	<350,000	
Leilenkopf	390,000, 405,000	K-Ar
Oldbruck	410,000	K-Ar
Hardt near Rieden	420,000	K-Ar
Schellkopf	570,000	K-Ar

Table 5.13. Analyses of potassic rocks of the East Eifel, West Germany

	1	2	3	4	5	6	7	8
SiO_2	43.6	43.2	43.6	43.8	43.5	44.0	43.7	44.4
TiO_2	2.8	2.8	2.8	2.6	2.7	2.7	2.7	2.1
Al_2O_3	13.3	13.2	13.4	13.5	13.3	13.9	13.6	15.6
Fe_2O_3	2.8	5.0	7.6	5.2	5.0	4.3	6.4	4.2
FeO	7.6	5.6	3.4	5.2	5.4	6.3	4.1	5.4
MnO	0.2	0.2	0.2	0.2	0.2	0.2	0.2	0.2
MgO	9.7	10.5	9.9	10.5	10.8	9.0	9.8	7.3
CaO	12.8	11.9	12.6	11.8	11.8	11.9	12.1	11.5
Na_2O	2.5	2.4	2.7	2.9	3.3	3.2	3.4	3.2
K_2O	3.0	3.0	3.1	3.4	3.4	3.5	3.4	4.1
H_2O^+	1.0	0.8	0.8	0.7	0.3	0.8	0.2	0.6
P_2O_5	0.5	0.5	0.5	0.5	0.5	0.5	0.5	0.2
Total	99.9	99.2	100.5	100.4	100.3	100.3	100.1	98.9
(ppm)								
Cr	230	210	230	240	210	135	165	–
Ni	110	150	130	140	155	95	105	–
Cu	60	60	55	55	55	60	55	–
Zr	235	255	245	230	235	280	240	–
Sr	730	795	760	725	755	890	745	–
Rb	85	90	80	80	80	85	80	–
Ba	900	860	870	730	900	660	910	–
Zn	65	60	55	60	60	70	60	–

1-7 Leucite basanites from the following localities: 1 and 3 are from Veitskopf; 2 is from Nickenicher Hummerich; 4 is from Nickenicher Sattel; 5-7 are respectively from Sattelberg, Rothenberg, and Kunkskopf. Analyses 1-6 are from Duda and Schminke (1978) and that of 7 and 8 are from Duda (1975). Rocks 1 to 5 also includes 0.1% CO_2. 8 is a leucite tephrite from Hochsimmer, East Eifel (Duda and Schminke 1978); it includes 0.1% CO_2

5.13 Leucite-Bearing Rocks of the Tristan da Cunha Islands

Baker et al. (1964) reported potassic rocks from the Tristan da Cunha Islands lying in the south Atlantic (30°05 south, 12°17 west) almost midway between South Africa and the east coast of South America. Tristan is located at the top of a composite volcanic cone, which rises about 2060 m above sea level and has long been considered one of the isolated volcanic islands that lie on or along flanks of the mid Atlantic ridge (Fig. 5.1). The following differentiation sequences was noted by Baker et al. (1964): alkali basalt → trachybasalt → trachyandesite → trachyte. Leucite is found in the trachybasalt (Table 5.14) in association with plagioclase, titanaugite, and a small amount of olivine and iron ore. Amphibole, which is frequently present, is sometimes completely resorbed. Plutonic xenoliths are abundant in the lava flows and consist of leucocratic to melanocratic types. The leucocratic type, which is rare, sometimes contains up to 90% feldspar. The melanocratic xenoliths consist of kaersutite, titanaugite, plagioclase, and iron ore in

Table 5.14. Analyses of leucite-bearing rocks from the Tristan da Cunha Islands and Utsuryo Island

	1	2	3	4	5
SiO_2	49.52	48.54	47.06	56.57	57.91
TiO_2	3.18	2.98	3.44	1.42	0.65
Al_2O_3	17.72	18.00	17.14	19.21	18.22
Fe_2O_3	2.55	3.78	3.29	1.90	1.90
FeO	5.66	5.18	6.65	3.31	3.20
MnO	0.18	0.18	0.18	0.10	0.27
MgO	3.42	3.32	4.35	1.37	1.01
CaO	7.58	8.49	9.00	3.57	3.58
Na_2O	4.98	4.74	4.08	5.80	4.93
K_2O	3.88	3.38	3.40	6.06	6.69
H_2O^+	0.29	0.14	0.37	0.15	0.82
H_2O^-	0.15	0.03	0.27	0.15	0.38
P_2O_5	1.09	1.18	0.75	0.67	0.48
ZrO_2	–	–	–	–	0.07
Total	100.20	99.94	99.98	100.28	100.11

1, 2, 3. Leucite trachybasalts from the Tristan da Cunha Islands (Baker et al. 1964)
4. Vulsinitic vicoite (tephritic leucite phonolite) from Arpong Hill, Utsuryo Island (Harumoto 1970)
5. Leucite trachyandesite from Arpong Hill, Utsuryo Island (Tsuboi 1920)

variable proportion. Le Maitre (1969) concluded that these xenoliths represent the accumulation of phases that crystallized from a trachy-basalt magma at depths between 5 and 25 km.

5.14 Leucite-Bearing Rocks of Manchuria, China

Ogura et al. (1936, 1938, 1939) described the petrography and chem-istry of highly potassic lavas from various localities of Manchuria, China, which comprise Wu-ta-lien chih, Erh-ko, and Chi-hsing shan.

5.14.1 Wu-ta-lien-chih

This volcanic district, located to the south of the Ussuri River (48°43' North and 126°7' East), consists of fourteen volcanic cones lying on the lava plateau. The fissure system in the Manchurian volcanic field in general is developed along northeast–southwest (China trend) and northwest–southeast (Korean trend) as a result of compressional and tensional forces (Tateiwa 1976). Pleistocene trachyandesite magma of Banak type (or banakite; Tomita 1967) was erupted along these fissures or their intersections in the Wu-ta-lien-chih district, forming an extensive lava plateau with an area of about 500 km² on the basement composed of granite, Cretaceous sandstone, and shale and Recent sediments. This was followed by successive eruptions of the same banakitic magma from the central vents, which formed very flat shield volcanoes. Later, twelve small volcanic cones ranging from 40 m to 160 m in relative height were built on this shield plateau. The cones are composed of scoria and lava flows of banakitic rocks with or without leucite, called "shihl-unite" by Ogura et al. (1936). More recently this activity was followed by the eruption of "shihlunite" lava flow in the central part of the district, covering an area of about 68 km². On this younger flow the new cones of Lao hei shan and Huo shao shan were formed during the eruption of 1720. They have nearly circular bases with a relative height of 166 m and 73 m, respectively.

Ogura et al. (1936) found that the plateau lavas are mainly leucite-free olivine basalt, very rich in potash, which was termed "banakite" by Tomita (1967) whereas shihlunite, which constitutes the Shih-lung lava flow and fourteen cones, often contains leucite as an important mineral. The rock contains microphyric olivine, and augite in a ground-mass consisting of olivine, augite, leucite, anorthoclase, plagioclase, magnetite, and glass (N = 1.545–1.550). The compositions of leucite-free and leucite-bearing rocks (Table 5.15) show little difference.

It is interesting to note that granitic blocks and quartz grains are often found as xenoliths or xenocrysts in the highly potassic rocks of this area. Gorai (1940) considered the formation of leucite-bearing rocks by the selective assimilation of mica and potash feldspar in the granitic rocks by a basaltic magma. The field observation of granitic xenoliths within potassium-rich lavas of Wu-ta-lien-chih may thus support his conclusion. The effectiveness of such a mechanism for the formation of leucite-bearing rocks will be discussed in a later section.

5.14.2 Erh-Ko

This volcano is located about 70 km south of the Wu-ta-lien-chih volcanic field (Ogura et al. 1938) and consists of three small cones: Tung

Table 5.15. Analyses of leucite-bearing rocks of Manchuria, China

	1	2	3	4	5	6
SiO_2	42.23	47.15	47.28	48.14	51.12	51.38
TiO_2	2.02	2.46	2.30	2.40	2.70	2.20
Al_2O_3	12.46	14.27	13.45	13.91	12.99	14.54
Fe_2O_3	2.89	9.32	7.00	9.10	4.36	8.09
FeO	10.93	1.44	3.43	2.00	5.39	0.85
MnO	0.19	0.18	0.08	0.09	0.06	0.07
MgO	11.81	7.78	8.44	8.86	6.01	6.34
CaO	10.30	8.32	7.72	7.22	5.84	6.16
Na_2O	3.91	3.25	4.56	3.31	3.24	3.72
K_2O	1.73	4.56	4.22	2.93	5.67	4.67
H_2O^+	1.50	0.79	0.77	0.83	1.02	0.64
H_2O^-	–	–	0.34	0.17	0.16	0.14
P_2O_5	0.71	1.07	0.75	1.01	0.98	1.13
Total	100.68	100.59	100.34	99.97	99.54	99.93

1. Leucite basanite from Nao-Po Shan, Ch'i-hsing Volcano (Ogura et al. 1939)
2. Leucite absarokite from older volcanic cone of Tung-lung-men Shan, Lung chiang Province (Ogura et al. 1936)
3. Leucite-olivine banakite from older volcanic cone of Mo-la-pu Shan, Lung chiang Province (Ogura et al. 1936)
4. Leucite absarokite from older volcanic cone of Tung-chao-te-pu Shan, Lung chiang Province (Ogura et al. 1936; cited by Tomita 1967)
5. Leucite absarokite from older volcanic cone of Yao-chuan Shan, Lung chiang Province (Ogura 1939; cited by Tomita 1967)
6. Leucite-olivine banakite from older volcanic cone of Pei-ko-la-ch'iu Shan, Lung-chiang Province (Ogura et al. 1939)

Shan, Hsiao-ko Shan, and Hsi Shan. Tung Shan is the largest cone above 100 m high and 700 m in diameter, whereas Hsi Shan is 75 m high and 500 m in diameter, lying on the basement of Cretaceous and Pleistocene sedimentary rocks.

Pleistocene volcanic activity started with the eruption of leucite basanite, which formed a shield-like plateau, covering an area of about 33 km². A second activity produced vesicular and scoriaceous lava flows of leucite basanite, which formed volcanic cones on the shield plateau. Chemical analyses of a leucite basanite from this area is given in Table 5.15.

5.14.3 Chi-hsing Shan

The Chi-hsing Shan volcano is located near the town of Ling-yuan (Cheng-Chia-Chun) in the center of the Manchurian plain (123°30' East, 43°30' North), and is built up of seven small volcanic cones (Ogura et al. 1939). Poli Shan is the highest cone, 110 m high, and Nao-Pao Shan is the lowest, 30 m high. The basement of the volcano consists of alternate beds of Pleistocene or Holocene sand and clay. The lava flows are mainly banakites but in some rare instances leucite basanite occurs at Nao-Pao Shan. The leucite basanite consists of olivine, augite, plagioclase, leucite, and magnetite, along with rare picotite and rhönite. Its chemical composition is given in Table 5.15.

5.15 Leucite-Bearing Rocks of Utsuryo Island (Ullung-do), Korea

The extinct volcano of Utsuryo (Ullung-do) Island (Fig. 5.1) is located in the western part of the Sea of Japan, 130 km off the eastern coast of Korea (37°30' North and 130°50' East). Though the height is about 983 m above sea level, the cone rises about 2200 m from the sea floor, where its basal diameter attains 30 km.

The first petrological study of the rocks of this island was made by Tsuboi (1920) and later by Harumoto (1930, 1970). The volcanic history of the island was summarized by Harumoto (1970) as follows:

1. The eruption of a vast quantity of basaltic lavas built up a huge submarine volcanic cone, more than 2000 m in height, which eventually formed an island. The cone was subsequently intruded by numerous basaltic dikes.

2. Paroxymal eruption of trachytic lavas formed a new pyroclastic cone, which was later intruded by swarms of trachytic dikes.
3. The third stage was characterized by the development of the sub-aerial body of the volcano on the ruined pyroclastic cone. Many trachytic lava flows were poured out through a central vent, followed by flank eruption of trachyte and phonolite.
4. These events were followed by the collapse of the summit and the northern flank of the cone, forming a caldera of about 3.5 km in diameter. Eruption of leucite-bearing lavas formed an intra-caldera dome at Arpong Hill.
5. The second paroxymal eruption took place at the central vent in the caldera, ejecting trachytic pumice and lapilli all over the island.

Basalt from this island has variable compositions ranging from a hypersthene- or pigeonite-bearing type to an olivine and titanaugite-bearing undersaturated variety. Trachytes usually contain soda-augite and fayalite, hastingsite, anorthoclase, or biotite. In some hastingsite-bearing trachytes analcite is also present. Soda-augite-bearing phonolites also occur as flows or dikes. Syenitic and monzonitic xenoliths, sometimes containing nepheline and sodalite, are frequently found in the trachytes.

Tsuboi (1920) first discovered leucite in the lavas from the central dome of Arpong Hill. He termed the rocks vulsinitic vicoite. Later Harumoto (1930, 1970) referred to them as trachyandesite. In the hand specimens of the leucite-bearing rocks, phenocrysts of feldspar, amphibole and augite can be observed. Anorthoclase occurs as microphenocrysts on thin peripheral rims around labradorite. Small leucite crystals, 0.05–0.25 mm in size are present in the groundmass. Two chemical analyses of the leucite-bearing rocks of the island are given in Table 5.14.

Chapter 6   Conditions of Formation of Leucite-Bearing Mafic and Ultramafic Rocks

6.1 Chemical Conditions

Formation of leucite in potassium-rich mafic and ultramafic rocks is dependent on the following special conditions:

1. Leucite forms in the lava with SiO_2, usually less than 52.50%, the ratio SiO_2/K_2O (wt.%) being lower than 20 (Washington 1906; Fig. 6.1).
2. Leucite may appear when the K_2O/Na_2O ratio is greater than 1. If the Na_2O content is higher than the K_2O content nepheline should appear, which can contain considerable amounts of kalsilite and some excess silica (Hamilton and MacKenzie 1960) in the form of leucite (to be discussed later) as solid solution, thus inhibiting the appearance of leucite.
3. The prohibitive tendency of high soda may be overcome by a low ratio of silica to potash (Washington 1906; p. 185).

6.2 Silica Activity in Leucite-Bearing Rocks

In a reaction of the type,

$$A + SiO_2_{(liquid)} = B \tag{1}$$

the silica activity as a function of ΔG^0 can be expressed by the following equations,

$$\Delta G_{reaction} = \Delta G^0_{reaction} - RT \ln a^{liquid}_{SiO_2} = 0 \tag{2}$$

$$\log a^{liquid}_{SiO_2} = \Delta G^0_{reaction}/2.303\ RT \tag{3}$$

where "a" denotes the activity and ΔG is the change in the free energy of formation. Carmichael et al. (1970) considered several sets of such

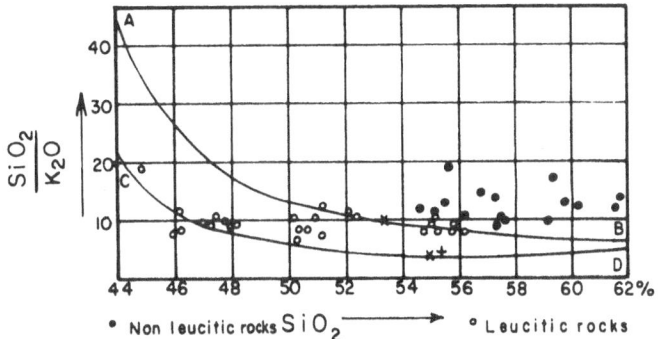

Fig. 6.1. Plot of leucite-bearing rocks in a diagram of SiO_2/K_2O vs SiO_2. (After Washington 1906)

reactions involving the silication of forsterite, nepheline, leucite, perovskite, forsterite + akermanite, forsterite + gehlenite, and larnite to form respectively enstatite, albite, orthoclase, sphene, diopside, diopside solid solution, and wollastonite. Most of these silica-undersaturated minerals are important mineral constituents of leucite-bearing rocks. Their study thus provides information regarding the silica activity in these rocks.

As a standard state of silica they used silica glass, rather than quartz or tridymite, since the mixing process would require disruption of bonds of the crystal lattice with other components. A natural silicate liquid can be regarded as a mixture of silica with other oxides. Therefore,

$$RT \ln a_{SiO_2} = \mu_{SiO_2}^{liquid} - \mu_{SiO_2}^0 \ , \ where \ \mu_{SiO_2}^{liquid} \tag{4}$$

where $\mu_{SiO_2}^{liquid}$ is the chemical potential of SiO_2 in liquid in the standard state. The energy difference of $(\mu - \mu^0)$ will be related to the mixing process.

Using the data of Robie and Waldbaum (1968) and Kelley (1960), Carmichael et al. (1970) determined the log $a_{SiO_2}^{liquid}$ values for different reactions as a function of temperatures. Their diagram is reproduced in Fig. 6.2. Leucite and albite do not coexist in nature under equilibrium condition. Orthorhombic pyroxene is also not an accompanying phase in these rocks. Larnite is also not observed in the leucite-bearing rocks, thus they lie between the nepheline and larnite silication curves (between v and iii, Fig. 6.2).

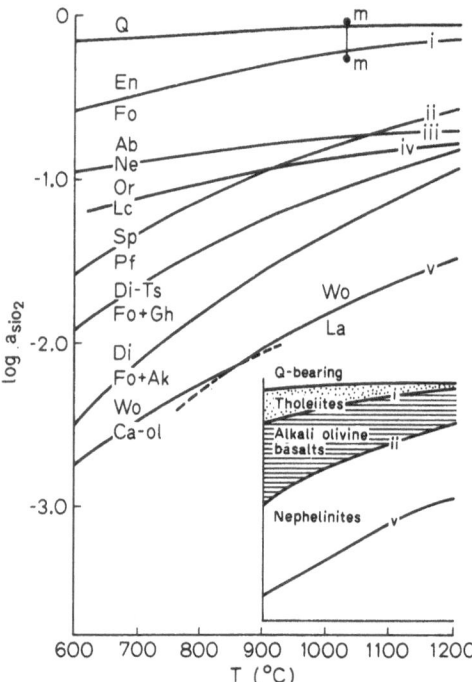

Fig. 6.2. Log $a_{SiO_2}^{liquid}$ vs temperature for various silication reactions. (After Carmichael 1967)

6.3 Physical Conditions

It is true that leucite is stable even above 19 kb dry pressure, as an incongruent melting phase of sanidine (Lindsley 1966; Fig. 14.6). However, the field of leucite is greatly reduced under water pressure, when it breaks down to K-feldspar and kalsilite. The field of leucite produced as an incongruent melting product also decreases (Tuttle and Bowen 1958; Fig. 14.2) with increasing water pressure, and finally it disappears (Morse 1968). Leucite stability under pressure will be discussed in more detail in a later section. However, the above discussion is conclusive enough to suggest that the presence of leucite will place some restriction on the physical and chemical condition of formation of leucite-bearing rocks. It has been found that pure akermanite also breaks down in the presence of water, at low temperatures and pressures to form wollastonite and monticellite (Harker and Tuttle 1956), whereas at variable temperatures under high pressures it breaks down to diopside and merwinite (Yoder 1968; Fig. 6.3). Breakdown of akermanite under CO_2 pressure has also been demonstrated by Yoder (1973). Forsterite

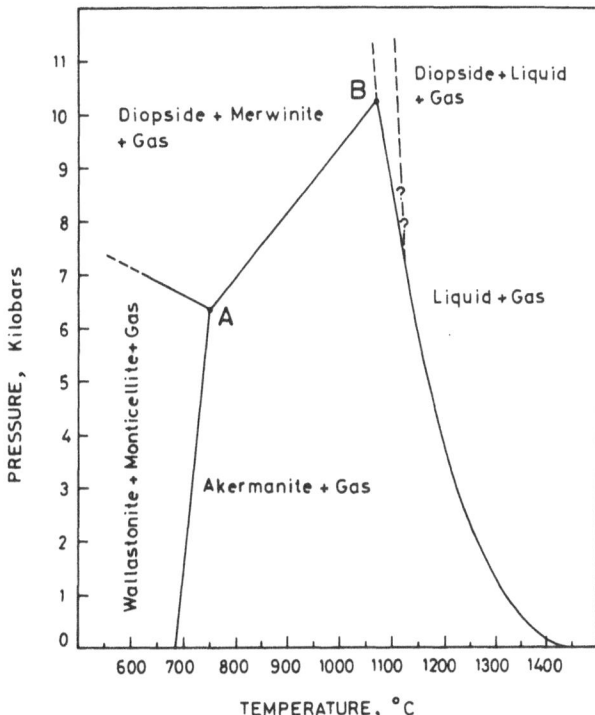

Fig. 6.3. Akermanite-H$_2$O system. The data between 2 and 4 kb are runs of Harker and Tuttle (1956). All other data are from Yoder (1968). (After Yoder 1973)

and akermanite together are also not stable at low temperatures and pressures. They react to form diopside and monticellite solid solution (Walter 1963; Yoder 1968). Willemse and Bensch (1964) provided field evidence of the low-temperature and low-pressure breakdown of aker-manitic melilite from the Bushveld complex. In this area they found graphic intergrowth of monticellite and wollastonite, indicating the breakdown of akermanite. Field evidence of the intergrowth of kalsilite and K-feldspar is also reported from the same area, along with an un-identified phase, probably leucite, which may indicate that these two phases represent the low-temperature and low-pressure breakdown of leucite.

The above discussion suggests that leucite, along with forsterite and akermanite, must be restricted to conditions of high temperatures and low pressures; thus leucite-bearing rocks are products of volcanic and subvolcanic activity.

Chapter 7   Leucite-Bearing Ternary Joins and Systems

7.1 The System Nepheline-Kalsilite-SiO$_2$

This is probably the most fundamental system related to the genesis of highly undersaturated alkalic rocks. Bowen (1937) considered that after fractional crystallization of a primary basalt magma the residual liquid will be highly enriched with respect to alkali, alumina, and silica and its bulk composition should lie in the nepheline-kalsilite-SiO$_2$ system. He thus called the system "petrogeny's residua system", which was investigated by Schairer and Bowen (1935) and Schairer (1950). Figure 7.1 shows that the join albite-sanidine divides the system into a silica-saturated and a silica-deficient portion. There are two temperature minima, one on the boundary between the tridymite and feldspar fields (G, Fig. 7.1) known as the granite minimum, and another one which lies in the silica-undersaturated region (M, Fig. 7.1), known as the nepheline syenite minimum. While the silica-rich portion of the system is important to understand the genesis of granitic rocks, the discussion here is mainly concerned with the silica-poor region of the system. Figure 7.1 shows that there is a large field of leucite$_{ss}$[*] which extends even beyond the albite-sanidine join. The field of nepheline$_{ss}$ covers a considerable portion of the system between the alkali feldspar and nepheline-kalsilite joins. The field of nepheline$_{ss}$ is separated from that of orthorhombic kalsilite$_{ss}$ at about 65% KAlSiO$_4$ when the former does not contain excess SiO$_2$, and at 80% when nepheline contains about 16% excess silica (Fig. 7.1). The course of crystallization of a melt can be best understood by considering liquids of three different compositions.

From a melt of initial composition 1, leucite$_{ss}$ should be the first phase to crystallize; then the composition of the liquid moves toward 2; where leucite$_{ss}$ begins to react with the liquid to crystallize alkali feldspar. The composition of the liquid moves along the boundary

---

[*] ss denotes solid solution. For other abbreviations see List of Abbreviations

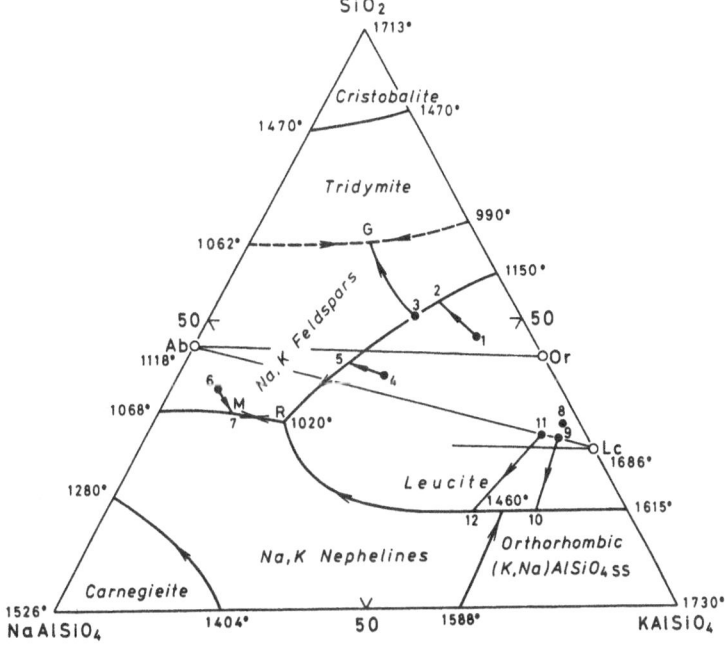

Fig. 7.1. Phase diagram of the system nepheline-kalsilite-SiO₂ at 1 atm. (After Schairer 1950). In the diagram various *points* and *lines* joining them are addition of the authors

curve to 3, where the resorption of leucite$_{ss}$ is complete and the liquid leaves the boundary to move along 3–G, crystallizing only alkali feldspar. At G, feldspar is joined by tridymite. If the liquid composition lies in the silica-deficient region of the alkali feldspar join (4, Fig. 7.1), leucite$_{ss}$ should again be the primary phase. With a drop in temperature, the composition of the liquid now moves along 4–5. At 5, feldspar should coexist with leucite$_{ss}$, the latter being continuously resorbed by reaction with the liquid. When the liquid reaches R, leucite$_{ss}$ and feldspar are joined by nepheline$_{ss}$ and their crystallization should continue until the liquid is exhausted. If a liquid of initial composition 6 is considered, alkali feldspar should be the first crystalline phase to appear. The liquid then moves along 6–7 to 7, where crystallization of both nepheline$_{ss}$ and alkali feldspar occurs. The liquid then moves along the boundary to M, where crystallization comes to an end by the simultaneous separation of nepheline$_{ss}$ and alkali feldspar. Comparison of the crystallization behaviors of liquids of compositions 1 and 6 explains the association of leucite$_{ss}$ but never nepheline$_{ss}$ in chilled groundmass, containing free silica. Bowen and Ellestad (1937) suggested that the formation of pseudoleucite (aggregates of nepheline and alkali feld-spar pseudomorph after leucite) can be understood from the course of crystallization of liquid of composition 4.

The silica-deficient portion of the system nepheline-kalsilite-SiO_2 under water pressure (Hamilton and MacKenzie 1965; Morse 1969; Taylor and MacKenzie 1975) will be discussed later.

7.2 The System Diopside-Leucite-SiO_2

Schairer and Bowen (1938) added diopside to the leucite-SiO_2 join of the nepheline-kalsilite-SiO_2 system to examine the effect of pyroxene on the incongruent melting of sanidine$_{ss}$. Their diagram is reproduced in Fig. 7.2. If the composition of a liquid lies at 1, leucite should crys-

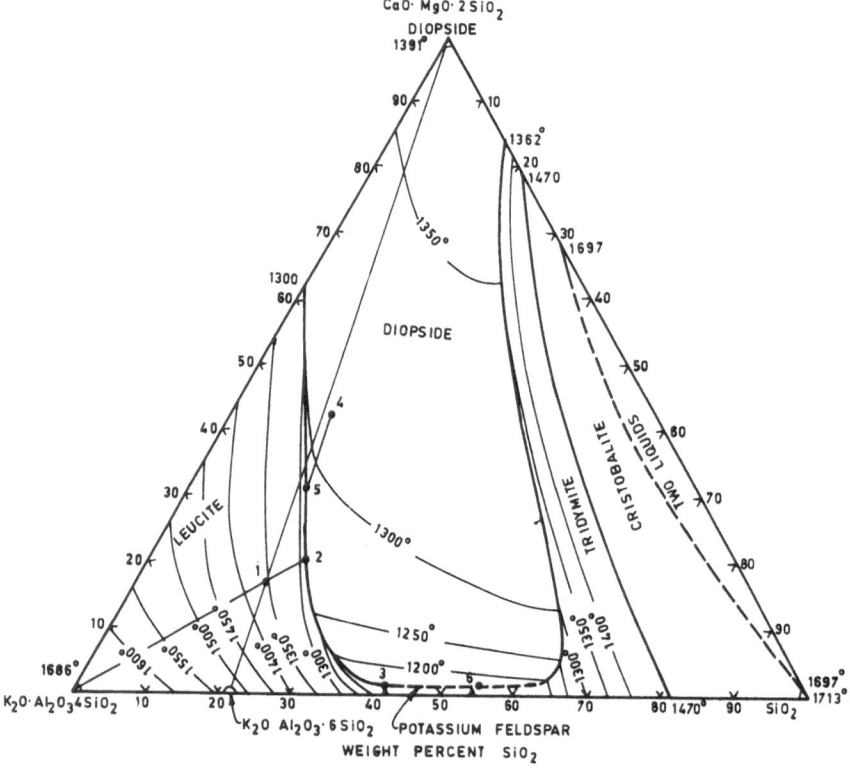

Fig. 7.2. Phase diagram of the system diopside-leucite-SiO_2 at 1 atm. (After Schairer and Bowen 1938)

tallize first; the composition of the liquid would move along 1-2. At 2 diopside$_{ss}$ should co-precipitate along with leucite$_{ss}$ and the liquid composition moves along 2-3 to 3, where these two crystalline phases are joined by sanidine$_{ss}$. The crystalline assemblages at 1, 2, and 3 would correspond to italite, leucitite, and pyroxene-leucite phonolite, respectively. With the drop of temperature, the liquid phase is eliminated during reaction at 3, although some leucite$_{ss}$ may still remain, if the original composition lies on the leucite side of the sanidine-diopside join.

From a liquid of composition 4, diopside$_{ss}$ crystallizes first as the liquid moves along 4-5. Both leucite$_{ss}$ and pyroxene crystallize at 5, while the liquid composition moves along 5-3 to 3, where sanidine$_{ss}$ starts to precipitate and the coexisting crystalline assemblage is diopside$_{ss}$ + leucite$_{ss}$ + sanidine$_{ss}$. As the temperature drops, leucite$_{ss}$ starts to react with the liquid and finally it disappears; the liquid then moves along 3-6. Tridymite, sanidine$_{ss}$, and pyroxene coexist with the liquid at 6 and the precipitation of these three phases should continue until the liquid is exhausted. The above discussion shows that from a leucitite magma (5), a quartz trachyte can be produced by equilibrium crystallization.

7.3 The System Anorthite-Leucite-SiO$_2$

This system (Fig. 7.3) was studied by Schairer and Bowen (1947). The course of crystallization of the liquid is very similar to that described in the preceding section except for the appearance of plagioclase instead of pyroxene. Point R where plagioclase, leucite$_{ss}$, and sanidine$_{ss}$ coexist is a reaction point, whereas point V is a eutectic where sanidine$_{ss}$, plagioclase, and tridymite are in equilibrium with the liquid.

7.4 The System Diopside-Sanidine-Albite

Morse (1969) added diopside to the albite-sanidine join (Fig. 7.4) of the system nepheline-kalsilite-SiO$_2$ to see the effect of the mafic components on the behavior of the alkali feldspar and feldspathoid-bearing melts. He found that the beginning of melting at 1 atm was close to 1020 °C, and is very similar to the minimum temperature of the nepheline-kalsilite-SiO$_2$ system (M, Fig. 7.1) under dry conditions.

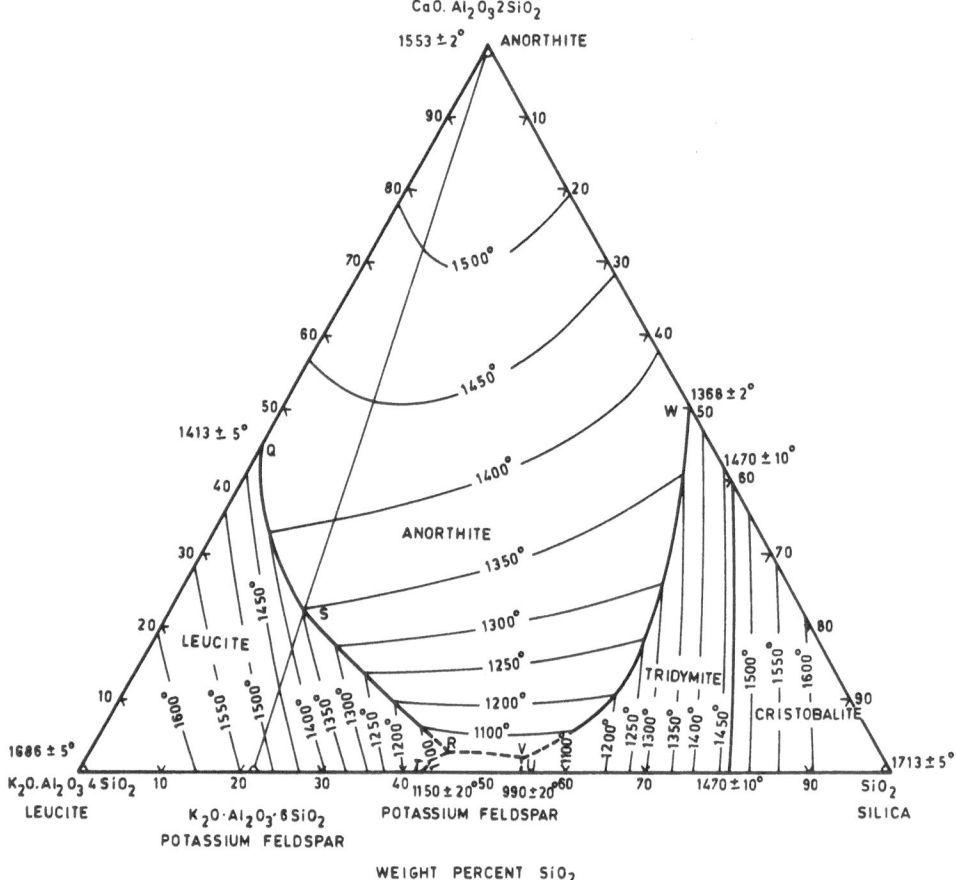

Fig. 7.3. Phase diagram of the system anorthite-leucite-SiO$_2$ at 1 atm. (After Schairer and Bowen 1947)

7.5 The System Diopside-Nepheline-Sanidine

The system (Fig. 7.5) of Platt and Edgar (1971) is important in establishing the paragenetic relationship between the rocks belonging to mela-nephelinite series and phonolites. They established that all phases in this system are solid solutions. If the presence of small amounts of possible Ca-Tschermak's molecule is ignored, the join can be treated as quinary, i.e., a part of the system CaO-MgO-NaAlSiO$_4$-KAlSiO$_4$-SiO$_2$. The join includes the primary phase fields of forsterite$_{ss}$, diopside$_{ss}$, nepheline$_{ss}$, and leucite$_{ss}$. Forsterite$_{ss}$ has a reaction relation with the

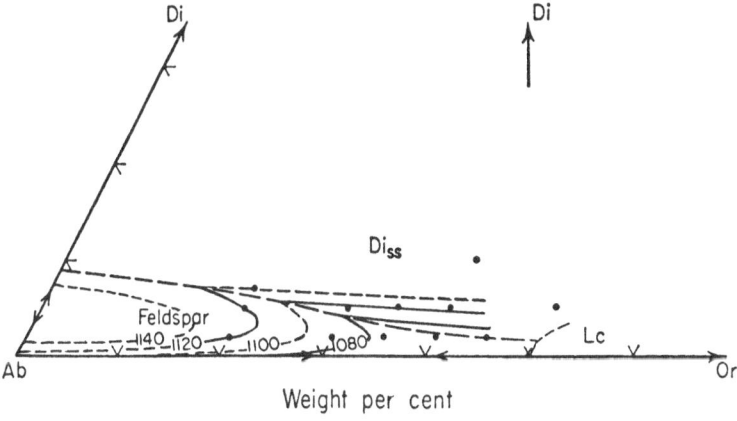

Fig. 7.4. Phase diagram of the system diopside-sanidine (Or)-albite at 1 atm. (After Morse 1969)

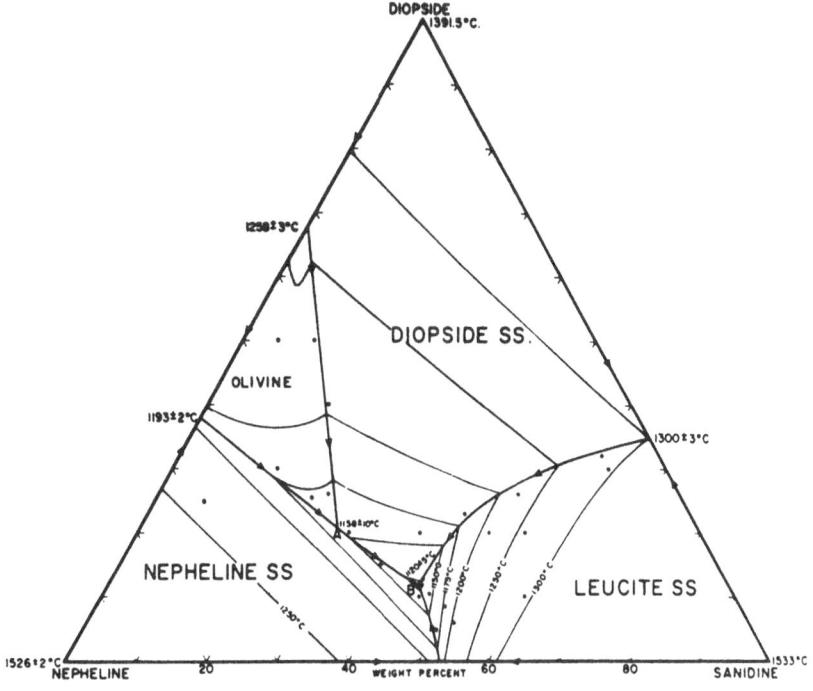

Fig. 7.5. Phase diagram of the system diopside-nepheline-sanidine at 1 atm. (After Platt and Edgar 1971)

liquid. The field of leucite$_{ss}$ is related to the incongruency of sanidine$_{ss}$, thus the former should react with the silica-rich liquid at low temperatures to form alkali feldspar. In the $KAlSi_3O_8$-poor part of the system melilite appears in addition to forsterite$_{ss}$. The solidus assemblages consist of the following phases: nepheline$_{ss}$ + diopside$_{ss}$ + leucite$_{ss}$ (1), nepheline$_{ss}$ + diopside$_{ss}$ + leucite$_{ss}$ + alkali feldspar (2), leucite$_{ss}$ + diopside$_{ss}$ + alkali feldspar (3), and nepheline$_{ss}$ + diopside$_{ss}$ + melilite (4). Assemblages (1), (2), (3), and (4) correspond to leucite nephelinite, nepheline-leucite phonolite, leucite phonolite, and melilite-nephelinite, respectively. The study of this system suggests that a pyroxene-rich nepheline phonolite can be produced from a nepheline leucitite melt, which is a derivative of a melilite-nephelinite magma. The last magma type itself is a product of an olivine-melilite-nephelinite magma. Their study thus supports the conclusions of King (1965), who suggested that nephelinites and phonolites from the Napak and Elgon volcanoes (eastern Uganda) were derived from mela-nephelinitic magma by early separation of forsterite, melilite, and pyroxene. It also supports the conclusion of Wright (1963), who considered that the differentiation of a mela-nephelinitic magma should explain the genesis of the Tertiary lava series in the western part of the rift system in Kenya. These lavas have variable compositions, which correspond to phonolites, nephelinites, and mela-nephelinites.

7.6 The System Nepheline-Leucite-Anorthite

Leucite and nepheline are the most commonly occurring feldspathoids in potassium-rich rocks. Whenever present in leucite-bearing rocks, plagioclase feldspars are always anorthite-rich (Shand 1943). The system nepheline-leucite-anorthite was investigated at one atmospheric pressure by Gupta and Edgar (1974) to determine the mutual phase relations of these minerals. Natural nepheline contains $CaAl_2Si_2O_8$, $KAlSiO_4$, and silica in solid solution (Miyashiro 1951; Donney et al. 1959). Similarly leucite incorporates sodium as the $NaAlSi_2O_6$ molecule in solid solution (Fudali 1963). The determination of the extent of solid solution may have important petrological implications.

The systems nepheline-anorthite (Gummer 1943) and leucite-anorthite (Schairer and Bowen 1947) have been published. The third join, nepheline-leucite, has been extrapolated from the system nepheline-kalsilite-silica (Schairer 1950).

Results given in Fig. 7.6 show that the nepheline-leucite-anorthite join cuts the primary phase volumes of β-alumina$_{ss}$, corundum
(a-Al_2O_3), carnegieite$_{ss}$, nepheline$_{ss}$, leucite$_{ss}$, and anorthite$_{ss}$, and is
thus a join of the quinary system Na_2O-K_2O-CaO-Al_2O_3-SiO_2. Four
four-phase assemblages occur at the following temperatures and compositions: (1) $1390° ± 5°C$, $Ne_{26}Lc_{22}An_{52}$; where β-alumina$_{ss}$, corundum, anorthite$_{ss}$, and liquid coexist; (2) $1285 ± 5°C$, $Ne_{49}Lc_{20}An_{31}$,
where nepheline$_{ss}$, β-alumina$_{ss}$, corundum, and liquid are in equilibrium;
(3) $1235 ± 5°C$, $Ne_{31}Lc_{44}An_{25}$; where corundum, nepheline$_{ss}$, and
anorthite$_{ss}$ coexist with liquid; (4) $1210 ± 5°C$, $Ne_{31}Lc_{44}An_{25}$; where
leucite$_{ss}$, nepheline$_{ss}$, anorthite$_{ss}$, and liquid are in equilibrium. The
five-phase assemblage of nepheline$_{ss}$, leucite$_{ss}$, anorthite$_{ss}$, corundum,
and liquid occurs at $1205 ± 5°C$.

Schairer and Bowen (1955) and Winchell and Winchell (1964) suggested that β-alumina is unstable in the absence of Na_2O and K_2O and
may exist as $Na_2O \cdot 6 Al_2O_3$. Semiquantitative electron microprobe

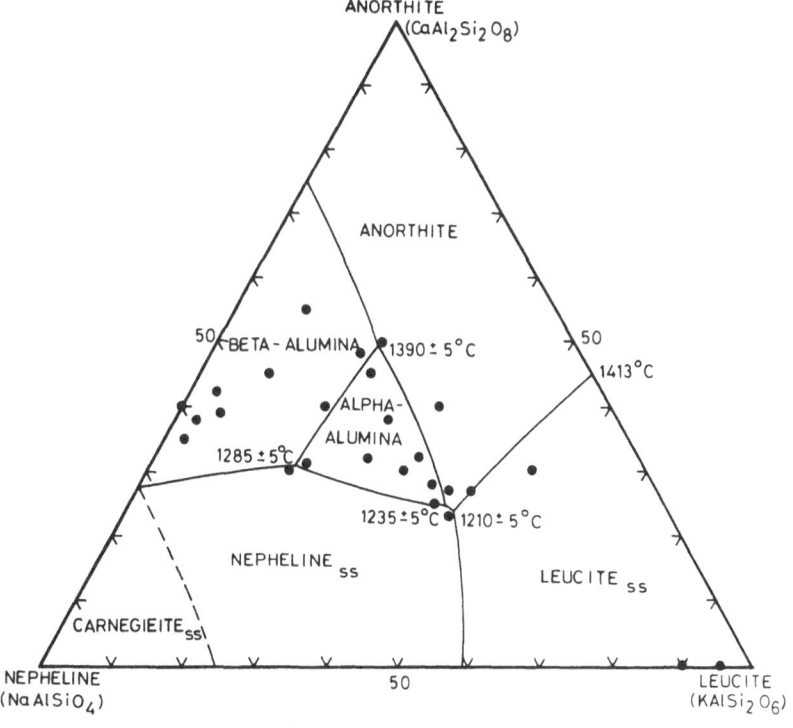

Fig. 7.6. Phase diagram of the join nepheline-anorthite-leucite at 1 atm. (After
Gupta and Edgar 1974)

analyses of β-alumina crystals (5–15 μ) indicate that they contain approximately 80 wt.% Al_2O_3. This suggests that β–Al_2O_3 exists here as $(K_2O \cdot Na_2O)$. 6 Al_2O_3 and has a reaction relationship with the liquid at temperatures ranging from 1340 °C for $Ne_{31}Lc_{21}An_{48}$ and $Ne_{27}Lc_{23}An_{50}$ to 1255 °C for $Ne_{50}Lc_{20}An_{30}$.

As a check that presence of corundum was not a result of metastability or loss of alkalis, two mixtures ($Ne_{32}An_{38}Lc_{30}$) and ($Ne_{34}An_{30}Lc_{36}$) lying within the primary phase fields of corundum were heated in sealed platinum capsules at 1300 °C for a week, and showed no change in the resulting assemblages.

The minute size of the crystals and high alkali content of the feldspathoids did not permit quantitative determination of their composition by microprobe. However, a reasonable estimate of their composition could be inferred from optical and X-ray diffraction methods.

In the system nepheline-anorthite, nepheline containing 23 wt.% $CaAl_2Si_2O_8$ is isotropic (Bowen 1912). In the studied portion of the system nepheline-leucite-anorthite, nepheline$_{ss}$ is nearly isotropic, suggesting that it contains amounts of anorthite comparable to that mentioned above. In all the mixtures with less than 20 wt.% $KAlSi_2O_6$, the final assemblage is nepheline$_{ss}$ + corundum + anorthite$_{ss}$ + liquid. Absence of leucite in these mixtures suggests that nepheline contains $KAlSiO_4$ and SiO_2 (in the combined form of leucite) in solid solution. Edgar (unpublished data) showed that $CaAl_2Si_2O_8$ does not affect the interplanar spacings of nepheline. The amounts of $KAlSiO_4$ and SiO_2 in nepheline$_{ss}$ from three mixtures were as follows: $Ks_{21.41}Qz_{3.91}$ ($Ne_{47}An_{31}Lc_{22}$), $Ks_{21.45}Qz_{3.90}$ ($Ne_{31}An_{23}Lc_{46}$), and $Ks_{21.56}Qz_{3.85}$ ($Ne_{16}An_{30}Lc_{54}$). The method of determination of the composition of nepheline is the same as described by Hamilton and MacKenzie (1960). This study therefore suggests that both potassium and calcium are equally favored in the nepheline structure. The composition of nepheline is thus controlled by the bulk composition of the melt from which it crystallizes, as suggested by Bowen and Ellestad (1937). Banister and Hey (1931) reported 4% CaO in natural nepheline, whereas Barth (1963) found 2% CaO.

Optical and X-ray study of two mixtures, $Lc_{95}Ne_5$ and $Lc_{90}Ne_{10}$, showed that below 1350 °C leucite contains more than 5 but less than 10 wt.% of nepheline.

The amount of anorthite incorporated by leucite in solid solution is very small, as suggested by the system leucite-anorthite (Schairer and Bowen 1947). The association of corundum with nepheline and plagioclase is usually found in metamorphic rocks and attributed to complex metamorphic reactions (Moyd 1949; Carlson 1957) or metasomatism (Gummer 1943). Thus the liquidus relations have little relevance to these rocks. However, the presence of corundum in diabase dikes in the Glen Riddle area of Pennsylvania (Tomilson 1939) may have a primary origin and has been produced by the type of reaction present in the synthetic system.

Leucite is normally absent in corundum-bearing undersaturated rocks. Under conditions of low to moderate water vapor pressure and in the presence of small amounts of MgO and FeO, leucite would react to form mica, a common accessory in such rocks. In leucite tephrites and basanites in which leucite$_{ss}$, Ca-rich plagioclase, and nepheline$_{ss}$ occur with ferromagnesian minerals, corundum is absent. In these rocks any excess alumina may combine with MgO and FeO to produce spinel ($MgAl_2O_4$) or hercynite ($FeAl_2O_4$).

Chapter 8 Leucite-Albite Incompatibility

It is well known that the assemblage of orthoclase and nepheline, but not that of leucite and albite, is possible in the calculation of the norm classification, based on petrological observations of leucite-bearing rocks (Cross et al. 1902). For example, Shand (1943) indicated that in leucite-bearing lavas the coexisting feldspar is either a calcium-rich plagioclase, or a potassium-rich alkali feldspar, or both. On thermochemical grounds, however, Miyashiro (1960) has postulated that the apparent incompatibility between leucite and albite may not exist at high temperatures. MacKenzie and Rahman (1968) described veins of leucite and sodium-rich feldspar in a basanite from the Massif Central, France, but they did not confirm this as a stable assemblage. Fyfe (quoted in MacKenzie and Rahman 1968) calculated that the leucite-albite assemblage may become stable below around 1000 °C. Experimental studies at P_{H_2O} up to 1 kb in the potassium-rich portion of the $NaAlSiO_4$–$KAlSiO_4$–SiO_2 system (Fudali 1963) indicate that the bulk compositions from which leucite-albite pair should crystallize have low melting temperatures and at low water pressures the end products are nepheline$_{ss}$, potassium-rich feldspar, and sodium-rich leucite. The only other study in this system is of the liquidus relations of the system nepheline-kalsilite-SiO_2 at atmospheric pressure (Schairer 1950; Chap. 7.1).

In an attempt to elucidate leucite-albite incompatibility and to determine the limits of plagioclase and alkali feldspar compositions, which may coexist with leucite in volcanic rocks, Gupta and Edgar (1975) studied the phase relations in the system leucite-albite and leucite-albite-anorthite at atmospheric pressure. The former represents a pseudobinary join in the Na_2O–K_2O–Al_2O_3–SiO_2 system in which subliquidus and solidus temperatures are sufficiently high to resolve the suggestion of Miyashiro (1960) and Fyfe (in MacKenzie and Rahman 1968) that leucite and sodium-rich feldspar may coexist at high temperatures. The leucite-albite-anorthite system is a pseudoternary join in the system, CaO–Al_2O_3–K_2O–Na_2O–SiO_2. The bounding joins leucite-anorthite and albite-anorthite in this system have been determined by Schairer and Bowen (1947) and Bowen (1913), respectively.

8.1 The Join Leucite-Albite

Figure 8.1 indicates that the join is pseudobinary, cutting the primary phase volumes of feldspar and leucite in the quaternary system Na_2O–K_2O–Al_2O_3–SiO_2. Point A ($Lc_{41}Ab_{59}$, 1068 ± 5 °C) represents a pseudoeutectic point, where leucite and feldspar coexist with liquid. The temperature is in good agreement with that of Schairer (1950) for the same chemical composition in the $NaAlSiO_4$–$KAlSiO_4$–SiO_2 system, where leucite, nepheline, and feldspar coexist with liquid. Point C (990° ± 5 °C) is the intersection of solidus in the system and the bound-

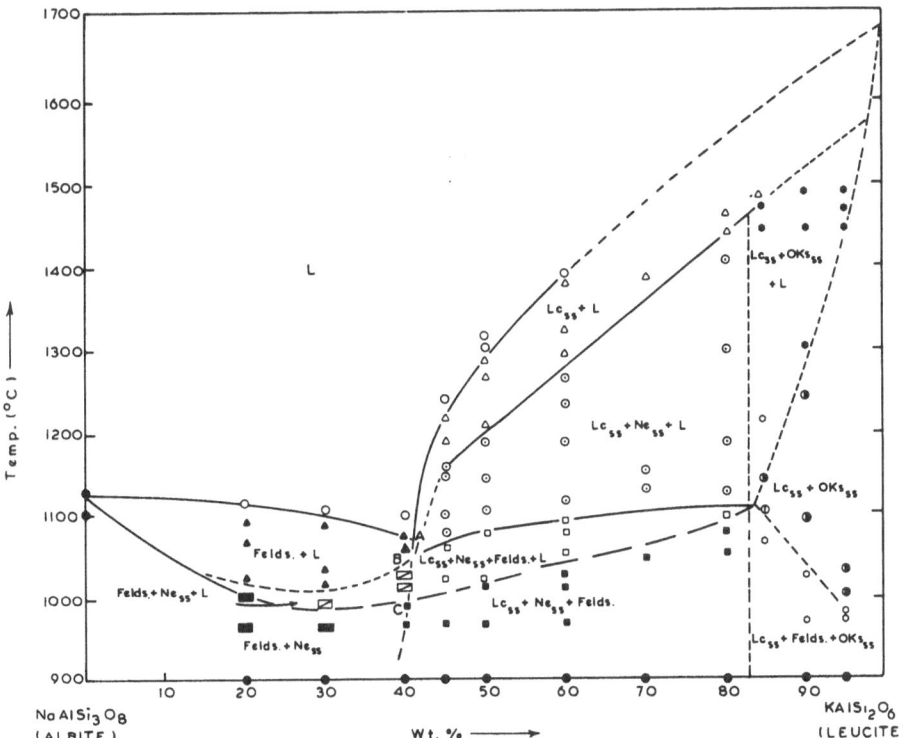

Fig. 8.1. Phase diagram of the join leucite-albite at 1 atm. (After Gupta and Edgar 1975)

ary curve of the field, where leucite and feldspar are mutually stable phases.

As shown in Fig. 8.1 the extent of solid solution of albite in leucite is very small, but increases with increasing temperature up to $Lc_{83}Ab_{17}$ and 1100 °C, where leucite$_{ss}$ coexist with orthorhombic kalsilite$_{ss}$. A slight increase in the interplanar spacings (211, 004, 400, and 420) of leucite, crystallized from bulk compositions $Lc_{95}Ab_5$ and $Lc_{90}Ab_{10}$ compared to those of its pure synthetic equivalent, and a decrease in those of (132 and 202) in kalsilites from the same compositions relative to pure kalsilite (Smith and Tuttle 1957), indicate that both these minerals are solid solutions. However, the exact compositions of these two phases were not determined.

The appearance of kalsilite$_{ss}$ and nepheline$_{ss}$ in the leucite-albite join can be explained if it is considered that leucite contains small amounts of excess silica and soda in solid solution, as is usually observed in nature. Its composition should thus lie along a compositional vector, which is a function of both SiO_2 and Na_2O, say at an arbitrary point 8 (Fig. 7.1). This point corresponds to leucite$_{ss}$ containing 6% $NaAlSi_3O_8$ and 5% SiO_2. If the crystallization of a liquid of bulk composition 9 is considered, with the drop in temperature, leucite$_{ss}$ of composition 8 should crystallize and the liquid should move along the line 9–10 towards the leucite$_{ss}$–orthorhombic kalsilite$_{ss}$ boundary and at 10, orthorhombic kalsilite$_{ss}$ should crystallize. If the bulk composition lies at 11, as the temperature decreases leucite$_{ss}$ should again appear as the primary phase and the composition of the liquid should move along the line 11–12 and at 12 nepheline$_{ss}$ should appear.

8.1.1 Composition of Alkali Feldspars in the Leucite-Albite-Join

In this join feldspars may coexist with leucite in the absence of liquid between $Ab_{59}Lc_{41}$ and $Ab_{17}Lc_{83}$ below about 1000° to 1100 °C depending on the bulk compositions (Fig. 8.1). Variations in cell parameters of six feldspars, crystallized at 960 °C from Ab_{100} to $Ab_{10}Lc_{90}$ are shown in Fig. 8.2a–g. Figure 8.2d shows a change in the slope of the angle a of the feldspars at about $Ab_{49}Lc_{51}$, indicating their change from triclinic to monoclinic symmetry.

Extrapolation of the until cell volume of the feldspars (Fig. 8.2g, Table 8.1), crystallizing from the composition $Ab_{59}Lc_{41}$, represents the most Na-rich feldspars that may coexist with leucite at subsolidus temperature in this join. With increasing K_2O in the bulk compositions and concomitant crystallization of leucite and eventually orthorhombic kalsilite, the rate of increase in the orthoclase component of the crystallizing feldspar decreases (Fig. 8.2g).

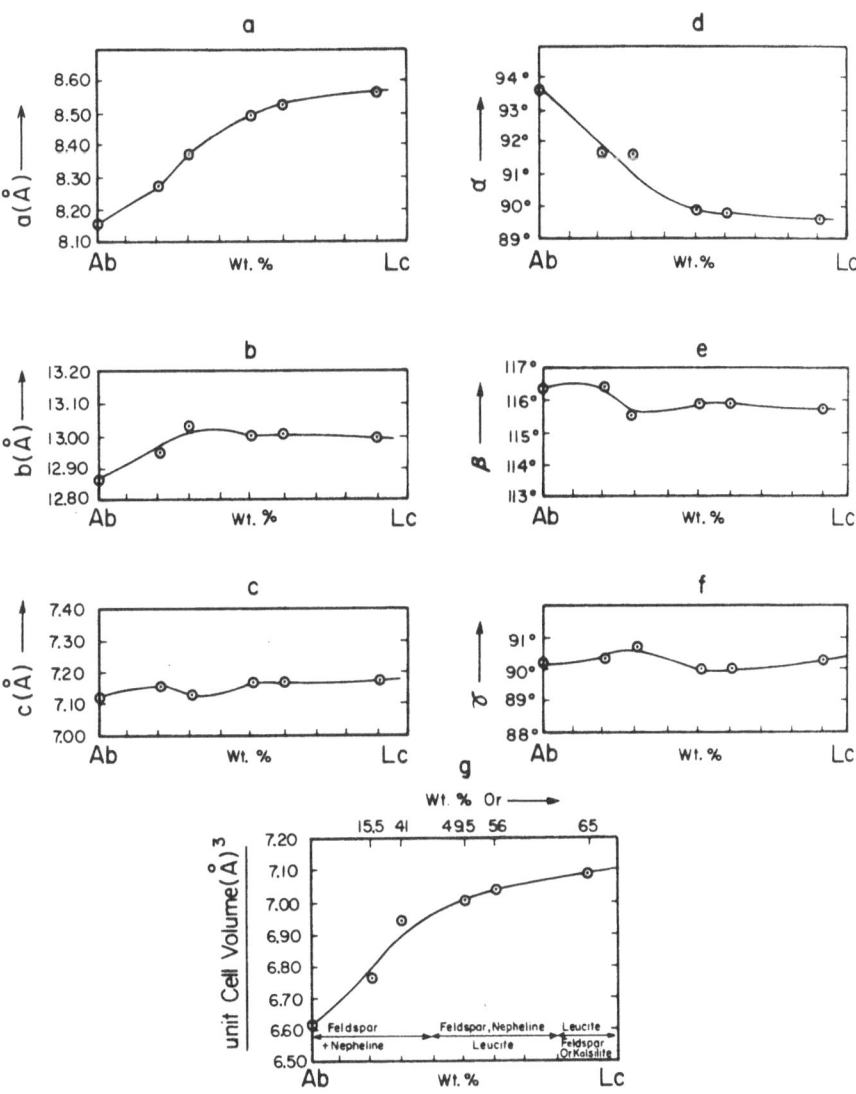

Fig. 8.2a–g. Cell parameters of feldspars in the join leucite-albite. (After Gupta and Edgar 1975)

Table 8.1. Composition of feldspars in the join leucite-albite at 1 atm and 900 °C

Bulk composition		Cell volume (Å)^3	Composition as wt.% determined from cell volume, following the method of Orville (1967)	
Ab	Lc		Ab	Or
100	0	661.51	100	–
80	20	677.19	84.5	15.5
70	30	694.49	59	41
50	50	700.19	50.5	49.5
40	60	704.26	44	56
10	90	709.05	35	65

Extrapolation of the composition of nepheline$_{ss}$ coexisting with feldspar at 960 °C from the mixture $Ab_{70}Lc_{30}$ gave $Ne_{68}Ks_{30}Qz_2$. When this composition is plotted along with coexisting feldspar composition ($Ab_{59}Or_{41}$) in the system $NaAlSiO_4 - KAlSiO_4 - SiO_2$, the tie line passes very close to the bulk composition $Ab_{70}Lc_{30}$, indicating the absence of any other phase. Compositions of nepheline$_{ss}$ crystallized at the same temperature from the bulk compositions $Ab_{60}Lc_{40}$ and $Ab_{50}Lc_{50}$ (where leucite is an additional phase) gave the same value ($Ne_{73}Ks_{25}Qz_3$). The composition of nepheline$_{ss}$ was determined by the method of Hamilton and MacKenzie (1960).

8.2 The Join Leucite-Albite-Anorthite

Phase relations in this pseudoternary join of the system $KAlSiO_4 - NaAlSiO_4 - CaAl_2Si_2O_8 - SiO_2$ are shown in Fig. 8.3. Only the liquidus relations within the system have been studied. However, it appears that from the bulk compositions lying close to the $KAlSi_2O_6 - CaAl_2 - Si_2O_8$ join leucite and plagioclase should be the only two crystalline phases. Figure 8.1, however, suggests that with the increase in the $NaAlSi_3O_8$ content of the bulk composition nepheline$_{ss}$ and alkali feldspar should be additional phases at low temperatures and when the starting material is considerably rich in the albite content, leucite should cease to appear.

Due to the pseudoternary nature of this join compositions of liquids do not lie in the leucite-albite-anorthite join. Estimation of the nature

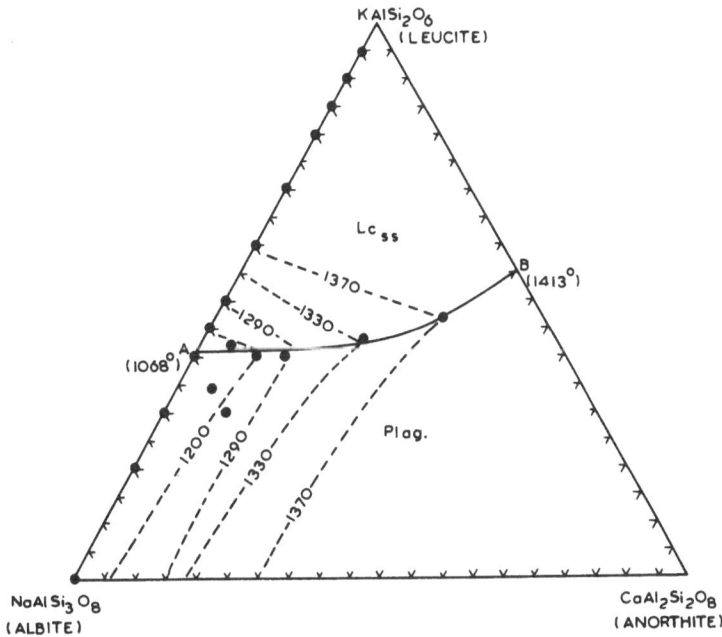

Fig. 8.3. Phase diagram of the join leucite-albite-anorthite at 1 atm. (After Gupta and Edgar 1975)

of these liquids can be made from the determination of the compositions of feldspars, crystallized at or near the boundary line A–B (Fig. 8.3).

8.2.1 Compositions of Ternary Feldspars in the Leucite-Albite-Anorthite Join

Microprobe analyses of feldspars were made on four bulk compositions, crystallized at temperatures close to liquidus, ranging from 1320 °C to 1180 °C (Table 8.2). For each feldspar the contents of CaO, Al_2O_3, and SiO_2 were determined by microprobe, assuming that $Na_2O + K_2O$ forms the remainder. In Fig. 8.4 tie lines have been drawn between the bulk compositions and the corresponding feldspar compositions projected on the leucite-albite-anorthite join. These tie lines show progressive enrichment in the albite-orthoclase contents of the feldspar with decreasing temperature, but indicate that the plagioclase, coexisting with a liquid low in anorthite content ($Lc_{53}Ab_{42}An_5$), is rich in anor-

Table 8.2. Microprobe analyses[a] of feldspars in the join leucite-albite-anorthite

Composition of mixtures			Temp (°C)	wt.%				An (wt.%)	Ab + Or (wt.%)
				SiO_2	$Na_2O + K_2O$	Al_2O_3	CaO		
Lc	Ab	An							
43	31	26	1,320	45.92	4.54	33.32	16.22	80.40	19.60
40	50	10	1,180	51.87	5.72	29.64	12.77	63.15	36.85
53	42	5	1,180	50.35	7.05	30.29	12.31	60.93	39.07
30	60	10	1,200	57.09	5.44	28.23	9.24	45.63	54.37

[a] CaO, Al_2O_3, and SiO_2 were determined; $Na_2O + K_2O$ was obtained by difference

thite (An_{61}, Ab–Or_{39}). This implies that the residual liquid after crystallization of such a feldspar must be enriched in $NaAlSi_3O_8$, although small amounts of soda may be incorporated in leucite. These results demonstrate that the addition of anorthite does not change the leucite-albite incompatibility.

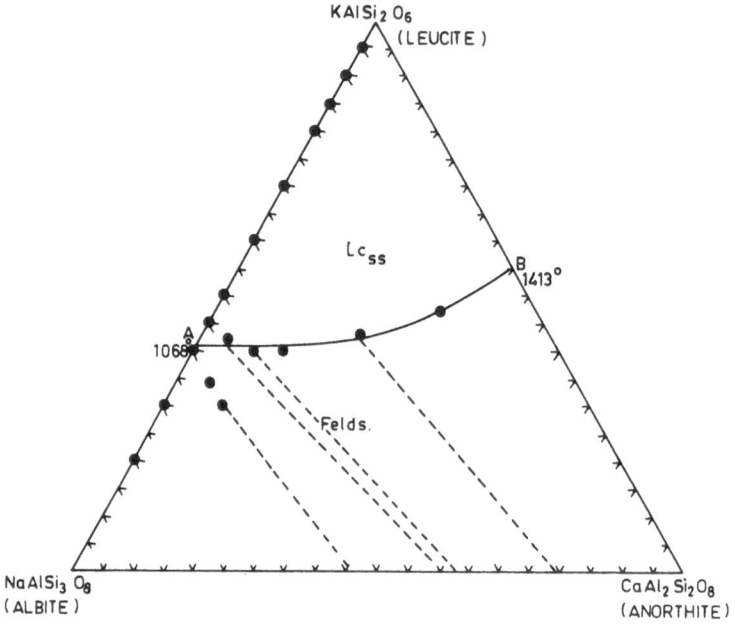

Fig. 8.4. Composition of plagioclases, coexisting with leucite in the system leucite-albite-anorthite. (After Gupta and Edgar 1975)

8.3 Petrological Implications

Study of phase relations in the join leucite-albite and leucite-albite-anorthite at atmospheric pressure supports conclusion of field studies that leucite and albite are incompatible. In the absence of the anorthite molecule the most sodium-rich feldspar that can coexist with leucite contains 46 wt. % orthoclase molecule. In the presence of the anorthite molecule, leucite coexists only with ternary feldspars, containing high proportion of anorthite molecule (approximately $An_{50}Ab-Or_{50}$). The incompatibility of leucite and albitic feldspar in the absence of the anorthite molecule is probably due to the following reaction:

$$NaAlSi_3O_8 + KAlSi_3O_8 \rightarrow (K, Na)AlSi_3O_8$$
albite   leucite   K-rich feldspar

$$+ (Na, K)AlSiO_4 + 2 SiO_2 .$$
nepheline   liquid

Thus residual liquids produced by such a reaction will be enriched in silica, although some of the silica may be incorporated in nepheline. In the leucite-albite-anorthite system, liquids closely representing natural magmas from which leucite-feldspar assemblages crystallize, the products of crystallization at low temperatures near the surface are anorthite-rich ternary feldspars, leucite, and a residual liquid, enriched in Na_2O and SiO_2. Some of the Na_2O may be incorporated into leucite. MacKenzie and Rahman (1968) noted that leucite rims in leucite-sodic feldspar veins in the Massif Central basanite became enriched in Na_2O with falling temperature. The presence of a residual liquid might also explain the albitic nature of the feldspar in these veins; the Na-rich feldspar having formed from an original K-rich feldspar by a process of alkali ion exchange. Such a mechanism would explain the maximum orthoclase content of 20 mol% (MacKenzie and Rahman 1968), relative to the albite-orthoclase content of 50 wt. % from the direct primary crystallization of low temperature liquids in the simplified leucite-albite-anorthite system.

The effects of P_{H_2O} on the incompatibility of leucite and albitic feldspar may be estimated by comparing the results of this study with that of Fudali (1963; Fig. 4) for the 850 °C isothermal section of the $NaAlSiO_4-KAlSiO_4-SiO_2$ at P_{H_2O} = 265 bars, where the stable assemblage for a bulk composition of approximately $Ab_{50}Lc_{50}$ is leucite$_{ss}$, nepheline$_{ss}$, and feldspar of composition about $Or_{60}Ab_{40}$. Table 8.1 shows that the corresponding bulk composition at atmospheric pressure crystallizes leucite$_{ss}$, nepheline$_{ss}$, and feldspars of the composition Or$_{49.5}$Ab$_{50.5}$. Thus crystallization under moderate P$_{H_2O}$ conditions appears to decrease the chances of compatibility of leucite and albite. Under higher P$_{H_2O}$ the stability of leucite-feldspar assemblages will be restricted to very potassium-rich compositions of feldspar.

The compositions of feldspars coexisting with leucites (Fig. 8.1) at temperatures above 1000 °C do not support the suggestions of Miyashiro (1960) and MacKenzie and Rahman (1968), who considered that leucite and Na-feldspar may be compatible at high temperatures.

Chapter 9 Leucite- and Feldspar-Bearing Quaternary Joins and Systems

9.1 The System Forsterite-Diopside-Leucite-Anorthite

The bulk compositions of plagioclase-bearing potassium-rich under-saturated rocks such as leucite tephrites and leucite basanites lie within the system forsterite-diopside-leucite-anorthite. These rocks occur in the Somma-Vesuvius region of Italy (Savelli 1967), the Birunga area of Uganda (Holmes and Harwood 1937; Ferguson and Cundari 1975), and the East Eifel region of West Germany (Duda and Schminke 1978). In these regions these rocks are often associated with phlogopite and/or olivine-bearing leucitites. In Fig. 2.3 compositions of some of these mafic and ultramafic rocks are plotted. A study of the system forsterite-diopside-leucite-anorthite may thus help to understand the paragenetic relationships between leucite-bearing tephrites and basanites and more mafic potassium-rich lavas without plagioclase.

The system forsterite-diopside-leucite-anorthite has four bounding joins: (1) forsterite-diopside-leucite, (2) forsterite-anorthite-diopside, (3) forsterite-leucite-anorthite, and (4) diopside-leucite-anorthite. Of these four joins, join (1) has already been discussed. The others are discussed below.

9.1.1 The Join Forsterite-Diopside-Anorthite. The join forsterite-diop-side-anorthite (Osborn and Tait 1952) has two piercing points: one at $Di_{22.5}An_{57.5}Fo_{20}$ and 1317 °C, the other at $Di_{49}An_{43.5}Fo_{7.5}$ and 1270 °C. At 1317 °C forsterite$_{ss}$, spinel$_{ss}$, and anorthite$_{ss}$ coexist with liquid, whereas at 1270 °C diopside$_{ss}$, forsterite$_{ss}$, and anorthite$_{ss}$ are in equilibrium with liquid and the assemblage at the end of crystalliza-tion consists of diopside$_{ss}$ + forsterite$_{ss}$ + anorthite$_{ss}$ + liquid.

9.1.2 The Join Forsterite-Anorthite-Leucite. This join has not yet been studied. However, results from the bounding joins forsterite-anorthite (Anderson 1915), forsterite-leucite (Schairer 1954), and leucite-anor-thite (Schairer and Bowen 1947) are already known. The phase diagram of this join can therefore be constructed as shown in Fig. 9.1. The spinel field in this case is small, judging from the join forsterite-anorthite-SiO_2 (Anderson 1915), and the association of leucite and Mg-spinel is not reported in nature, although the latter occurs in xenoliths within potassic lavas of East Eifel (Duda and Schminke 1978).

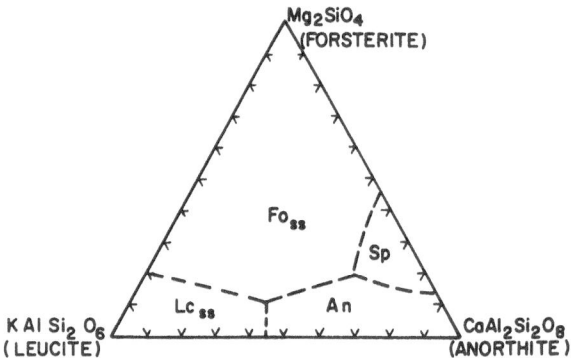

Fig. 9.1. Hypothetical phase diagram of the join forsterite-anorthite-leucite

9.1.3 The Join Diopside-Leucite-Anorthite. Experimental results on this join are summarized in Fig. 9.2. At A ($Di_{43}An_{35}Lc_{22}$), diopside$_{ss}$ + leucite$_{ss}$ + anorthite$_{ss}$ coexist with liquid. Optical and X-ray diffraction studies indicate that diopside may incorporate Ca-Tschermak's molecule whereas leucite probably includes $K_2O \cdot MgO \cdot 5 SiO_2$ in solid

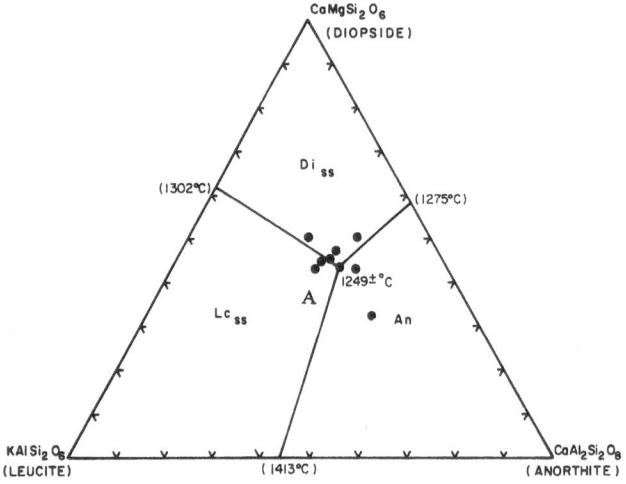

Fig. 9.2. Phase diagram of the join diopside-leucite-anorthite at 1 atm. (After Gupta unpublished)

solution. The composition of the liquid thus does not lie in this join, and point A therefore is a piercing point (1249° ± 5 °C).

9.1.4 Paragenesis. On the basis of the phase diagrams of these four bounding joins, the course of crystallization of a liquid in the system forsterite-diopside-leucite-anorthite, can be described by a simplified flow sheet diagram as given in Fig. 9.3, in which a rock nomenclature corresponding to each assemblage is also shown. This shows that a leucite-bearing basanite can be produced from an olivine leucitite (m, Fig. 9.3), a tephrite (n), an olivine plagioclase italite (p), or a liquid (o), approaching the composition of a magma, belonging to the shoshonite-absarokite series. Lavas of this type have been reported by Holmes and Harwood (1937; p. 145) from the Kogoma area of Birunga.

In the Bufumbira area of Uganda (Holmes and Harwood 1937; p. 98) leucite basanite locally grades into olivine-poor leucite tephrite. In the Mabungo area Holmes and Harwood (1937) found inclusions of leucite tephrites in leucite basanites. Reference to Fig. 9.3 shows the paragenetic relationship between these two rock types. Lacroix (1917)

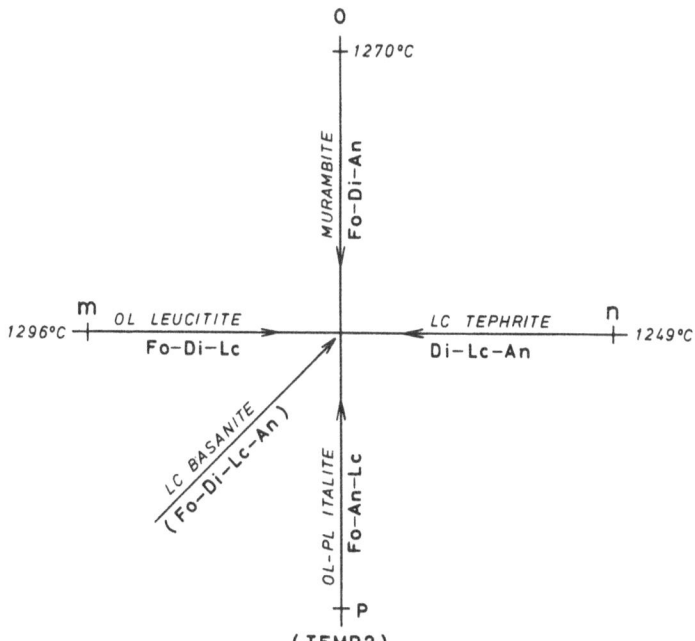

Fig. 9.3. Flowsheet and rock-nomenclature diagram of the system forsterite-diopside-leucite-anorthite

described leucite tephrites from the Kitale area, about 24 km to the west of Bufumbira, where leucite tephrites are found in association with leucitite. Figure 9.2 suggests that if the composition of the liquid lies close to the diopside-leucite join inside the system diopside-leucite-anorthite, leucitite magma will produce a leucite tephrite. Holmes and Harwood (1937) described lava flows of leucite basanite from the following areas of the Birunga (Chap. 5.2) petrographic province: Busumba, Mihanga, Mabungo, Kumizika, Ishozi, Nyakimanga, and Lugandable. Lavas transitional between olivine leucitite and leucite basanite (Fig. 9.3) have been described by Finckh (1912; p. 8) from the Kisi area near the northeast corner of Lake Kivu in Uganda.

Many older volcanic districts to the northwest of Vesuvius are characterized by leucite tephrites. In the Vulsinian district the lavas change from leucite tephrite to leucitite and very basic basanites (Washington 1906). In some places the latter rock types are low in modal plagioclase and are similar to olivine leucitite.

Duda (1975) found that leucite basanites crop out at Kunkskopf, Veitskopf, Rothenberg, Nickenicher Hummerich, Nickenicher Sattel, Heidskopf, and Sattelberg in the East Eifel region of West Germany (Fig. 5.17). In Kunkskopf, the association of leucite tephrite and basanite is found. Leucite tephrite was also described from Alteberg, E. Eifel by Duda.

9.2 The System Diopside-Leucite-Anorthite-SiO$_2$

The paragenetic relationship between leucite tephrite, phonolite, trachyte, and potassium-rich latite can be understood from the study of the system diopside-leucite-anorthite-SiO$_2$. Experimental results on various joins of the system are already known. The piercing points, mineralogical assemblages and their temperatures are summarized in Table 9.1. The system diopside-leucite-anorthite-SiO$_2$ has four bounding joins (Fig. 9.4): diopside-leucite-anorthite, diopside-leucite-SiO$_2$, anorthite-leucite-SiO$_2$, and diopside-anorthite-SiO$_2$. The join diopside-anorthite-orthoclase (sanidine) divides the system into a silica-saturated and a silica-undersaturated portion. On the basis of the results of these ternary joins, the various phase volumes and the courses of crystallization of melts within the system can be constructed as shown in Fig. 9.4. The system has two eutectic points, E_1 (silica-undersaturated) and E_2 (silica-saturated). If the initial composition of the liquid lies in the silica-deficient portion of the system, the composition of the melt

Table 9.1. Temperatures of various piercing points within the system diopside-leucite-anorthite-SiO$_2$

Piercing point	Temperature (°C)	System	References
1	1,249	Diopside-leucite-anorthite	Gupta (unpublished)
2	<1,200	Diopside-leucite-SiO$_2$	Schairer and Bowen (1938)
3	?	Diopside-sanidine-anorthite	Not determined
4	<1,100	Anorthite-leucite-SiO$_2$	Schairer and Bowen (1947)
5	<1,200	Diopside-leucite-SiO$_2$	Schairer and Bowen (1938)
6	< 990	Anorthite-leucite-SiO$_2$	Schairer and Bowen (1947)
7	1,200	Diopside-anorthite-SiO$_2$	Hytonen and Schairer (1961)

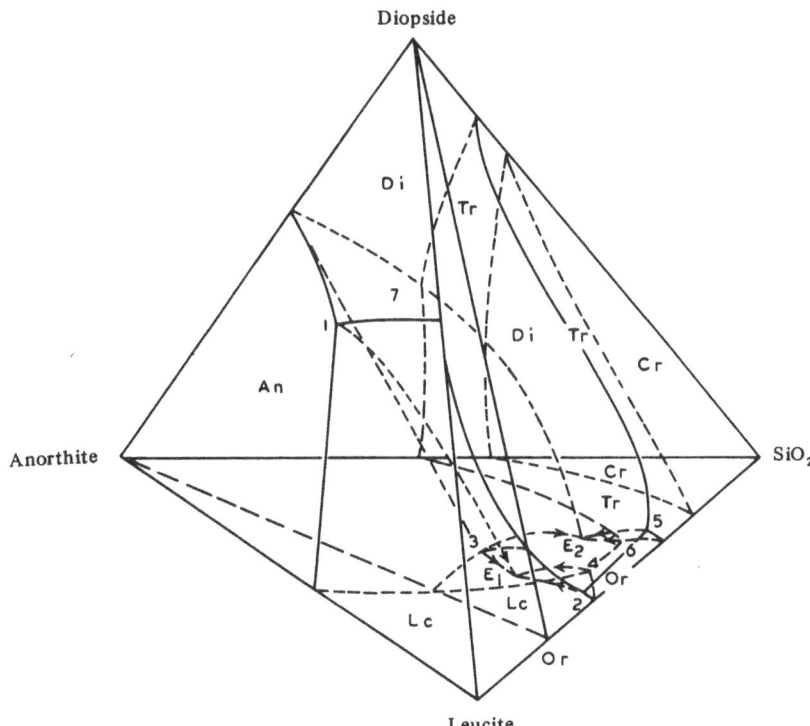

Fig. 9.4. Course of crystallization of liquid in the system diopside-leucite-anorthite-SiO$_2$ at 1 atm

should move towards E_1, whereas the liquid in the silica-saturated portion leads towards E_2. The coexisting assemblages at the eutectic points are diopside + leucite + sanidine + anorthite + liquid (E_1), and diopside + sanidine + plagioclase + tridymite + liquid (E_2). The univariant assemblages of diopside-anorthite-leucite-liquid ($1-E_1$, leucite tephrite), diopside-sanidine-leucite-liquid ($2-E_1$, pyroxene leucite phonolite), diopside-sanidine-anorthite-liquid ($3-E_1$, shoshonite), and anorthite-leucite-sanidine-liquid ($4-E_1$, plagioclase-leucite phonolite) lead to E_1. The univariant assemblages of diopside-sanidine-tridymite-liquid ($5-E_2$), anorthite-sanidine-tridymite-liquid ($6-E_2$), diopside-anorthite-tridymite-liquid ($7-E_2$), and diopside-sanidine-anorthite-liquid ($3-E_2$) move toward E_2. It should be noted here that at point 3, which lies in the diopside-anorthite-sanidine (Or) join, diopside, leucite, anorthite, and liquid coexist. However, because of reaction with the liquid, leucite should disappear and the resulting assemblage thus consists of diopside + anorthite + sanidine + liquid.

The univariant assemblages in the silica-saturated portion of the system and the coexisting phases at E_2 correspond to the trachyandesite-shoshonite-quartz latite series. Potassium-rich rocks consisting of various combinations of leucite, K-feldspar, clinopyroxene, and plagioclase have been reported from the Somma-Vesuvius (Savelli 1967; Fig. 2.3) and the Vico area (Cundari and Mattias 1974) of Italy and the Bufumbira region of East Africa (Ferguson and Cundari 1975). Leucite and nepheline-bearing tephrites have been found at Rothenberg (43; Fig. 5.18), Epfelsberg (68; Fig. 5.18), and Krufterofen (67; Fig. 5.18) in the East Eifel region by Duda (1975).

When Cundari and Mattias (1974) plotted the mineralogical compositions of the rocks from the Vico area of Italy in a triangular diagram of alkali feldspar (A)-plagioclase (P) and feldspathoid (F), most of the rocks plotted in the tephrite-leucite phonolite (TLP) field of Fig. 5.11. This figure shows that under equilibrium conditions of crystallization the final product would be a tephritic-leucite phonolite. Cundari and Mattias (1974) observed a general gradation from tephritic leucitite (TL) through phonolitic leucite tephrite (PLT) to alkali trachyte (LTR), and concluded that the parent magma had its composition in the TL–PLT–TLP field.

Ferguson and Cundari (1975) found that there are two series of the leucite-bearing rocks in the Bufumbira area of East Africa in an APF diagram (Fig. 5.4), i.e., they underwent the following paragenetic sequences:

Series A: leucitic tephrite, tephritic leucitite, leucitite, phonolitic leu-
          citite, and phonolite.

Series B: leucitic tephrite, phonolitic leucite tephrite, tephritic phono-
          lite, latite, and trachyte.

In their study of the volcanism in the Eolian arc region, Barberi et
al. (1974) found a close relationship between leucite tephrite and shoshonite in Vulcanello, of which the former rock type was probably
produced from the latter by fractionation.

Highly vesicular lava flows of leucite and K-feldspar-bearing tephrites
(E$_1$, Fig. 9.4), covering extensive areas have been reported by Holmes
and Harwood (1937) from the Sagitwe region of Birunga. Leucite trachyte has been described from the Karsimbi area of the Birunga region
by the same workers.

If the composition of the liquid lies in the silica-rich side of the join
diopside-orthoclase-anorthite, the final product would be quartz latite
(E$_2$, Fig. 9.4). Holmes and Harwood (1937) described rock types of
trachyandesite-latite series from the Sabyno area of Birunga volcanic
province, where the prominent rock types are potassium-rich mafic and
ultramafic varieties.

Chapter 10 The System
Forsterite-Diopside-Akermanite-Leucite

The four essential minerals of many potassium-rich mafic and ultra-mafic rocks are leucite, augite, olivine, and melilite. The bulk composition of these rocks thus lies within the phase volume of forsterite-diopside-akermanite-leucite (Fig. 2.1). A study of the system is thus applicable to the understanding of the phase relations and origin of potassium-rich mafic and ultramafic volcanic rocks. A detailed study of the system was made by Gupta (1972), who showed that all phases in this system are solid solutions. Forsterite may contain monticellite; diopside, and akermanite possibly incorporate Ca-Tschermak's molecule and gehlenite respectively; leucite may include $K_2O \cdot MgO \cdot 5\,SiO_2$ and monticellite probably contains forsterite. The four joins which bound the system are described below.

10.1 The Join Diopside-Akermanite-Leucite

Phase equilibrium diagram of the join is presented in Fig. 10.1. At point B ($Di_{39}Ak_{29}Lc_{32}$), diopside$_{ss}$ + akermanite$_{ss}$ + leucite$_{ss}$ + liquid coexist in equlibrium at $1281° \pm 3\,°C$. In their study of the thermal behavior of pure synthetic akermanite, Schairer et al. (1967) found well-distributed inclusions of diopside$_{ss}$ and wollastonite$_{ss}$ within melilite at and below $1240° \pm 3\,°C$. Wollastonite$_{ss}$ is absent on the liquidus of the present join. As the compositions of the crystalline phases do not lie within the join, the system is not ternary and point B is therefore a piercing point and the lines dividing the primary phase fields are traces of divariant surfaces cut by the present join.

10.2 The Join Forsterite-Diopside-Leucite

Figure 10.2 shows the equilibrium diagram for the join forsterite-diopside-leucite. At point c ($Fo_3Di_{60}Lc_{37}$) forsterite$_{ss}$ + diopside$_{ss}$ + leu-

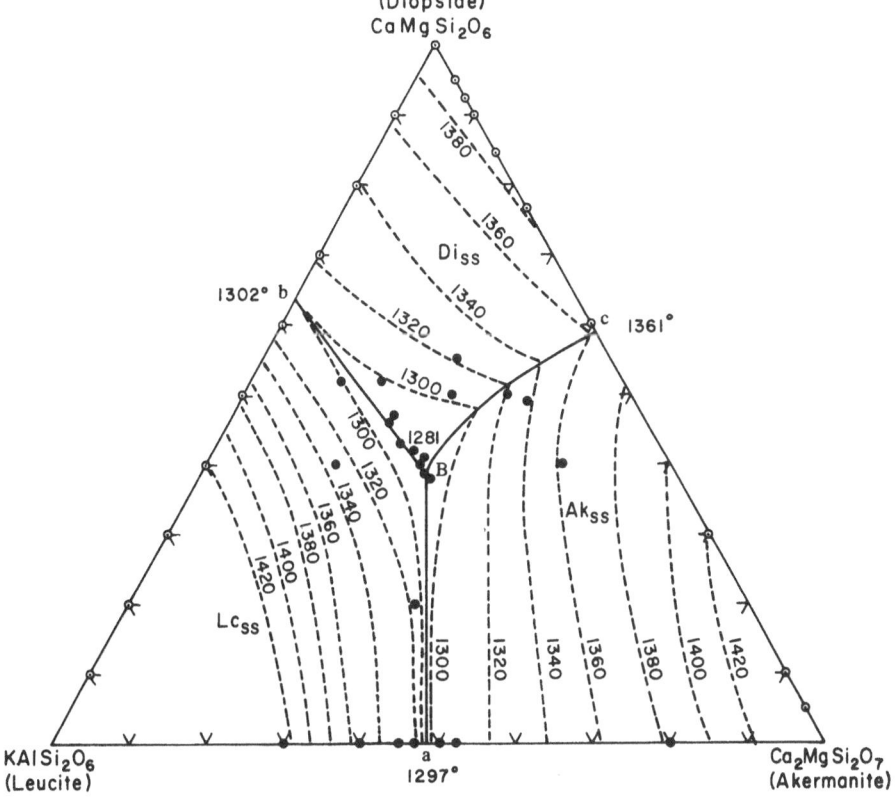

Fig. 10.1. Phase diagram of the join diopside-akermanite-leucite at 1 atm. (After Gupta 1972)

cite$_{ss}$ + liquid coexist in equilibrium at 1296° ± 3 °C. Again the system is not ternary and point c is a piercing point

10.3 The Join Forsterite-Akermanite-Leucite

Phase diagram of the join is presented in Fig. 10.3, which shows that it is pseudoternary and cuts through the phase volume of monticellite$_{ss}$. The join has two piercing points at Fo$_{17}$Ak$_{78}$Lc$_5$ and 1428° ± 3 °C (H), where the assemblage is forsterite$_{ss}$ + akermanite$_{ss}$ + monticellite$_{ss}$ + liquid; and at Fo$_9$Ak$_{43.5}$Lc$_{47.5}$, and 1286° ± 3 °C (G) the assemblage is forsterite$_{ss}$ + akermanite$_{ss}$ + leucite$_{ss}$ + liquid. At H, monticellite$_{ss}$ reacts

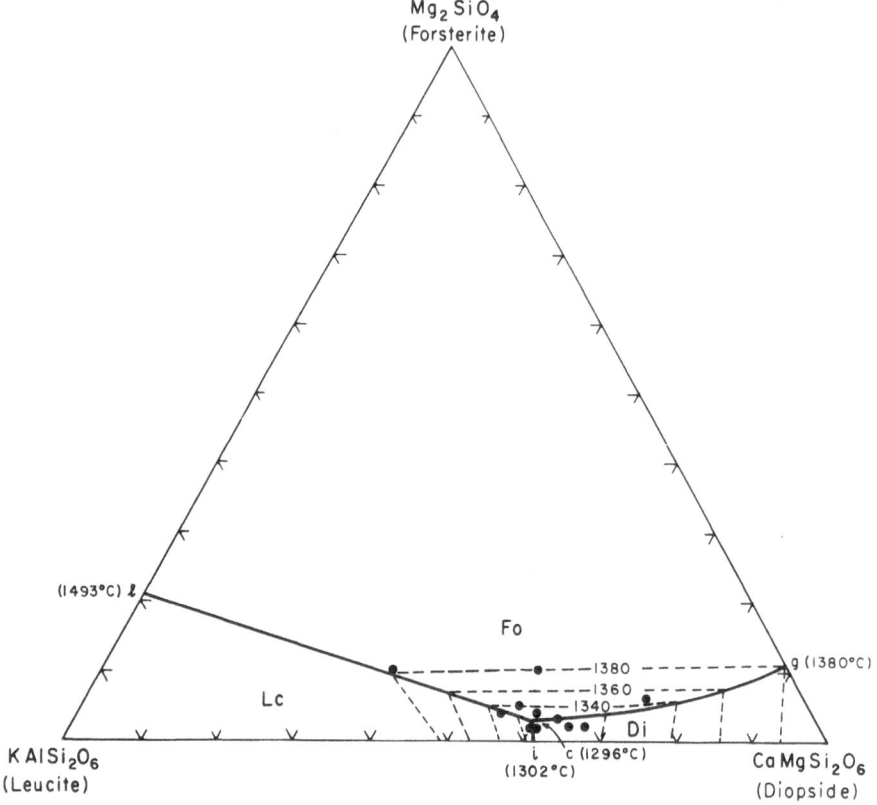

Fig. 10.2. Phase diagram of the join forsterite-diopside-leucite at 1 atm. (After Gupta 1972)

with the liquid to form akermanite and is completely eliminated at 1410° ± 3 °C. Therefore monticellite$_{ss}$ does not coexist with leucite$_{ss}$.

10.4 The Join Forsterite-Diopside-Akermanite

This join was studied by Ferguson and Merwin (1919) as a part of the system CaO–MgO–SiO$_2$. They found that at Fo$_{17}$Ak$_{77}$Di$_6$ and 1430 °C (estimated from their diagram) forsterite$_{ss}$ + monticellite$_{ss}$ + akermanite$_{ss}$ + liquid are in equilibrium. Another piercing point occurs at Fo$_{8.5}$Di$_{50}$Ak$_{41.5}$, where forsterite$_{ss}$ + diopside$_{ss}$ + akermanite$_{ss}$ + liquid coexist at 1357 °C. In this join monticellite$_{ss}$ reacts with the liquid to

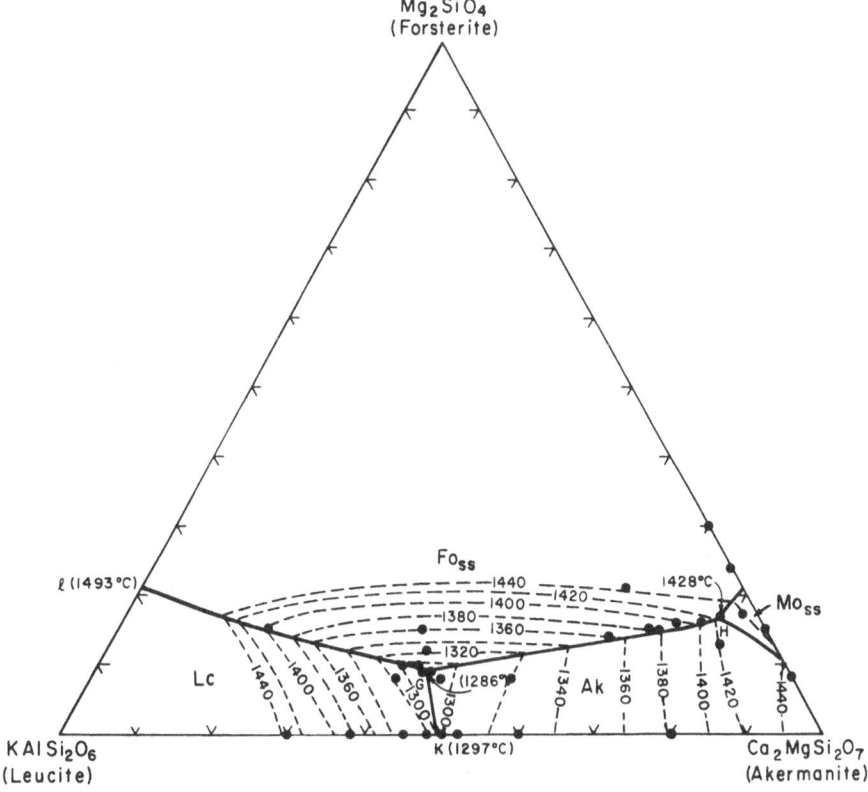

Fig. 10.3. Phase diagram of the join forsterite-akermanite-leucite at 1 atm. (After Gupta 1972)

form akermanite$_{ss}$ and the final crystalline assemblage is forsterite$_{ss}$ + diopside$_{ss}$ + akermanite$_{ss}$.

10.5 The Join Diopside-Akermanite-Leucite with 3% Forsterite

Study of the bounding joins of the tetrahedron suggests that the invariant point, where forsterite$_{ss}$ + akermanite$_{ss}$ + diopside$_{ss}$ + leucite$_{ss}$ + liquid are in equilibrium, lies very close to the basal plane diopside-akermanite-leucite, probably containing less than 5% forsterite. Accordingly, the join diopside-akermanite-leucite with 3% forsterite was studied to determine the temperature of the invariant point. The equilibrium diagram for the join is represented by Fig. 10.4. At K

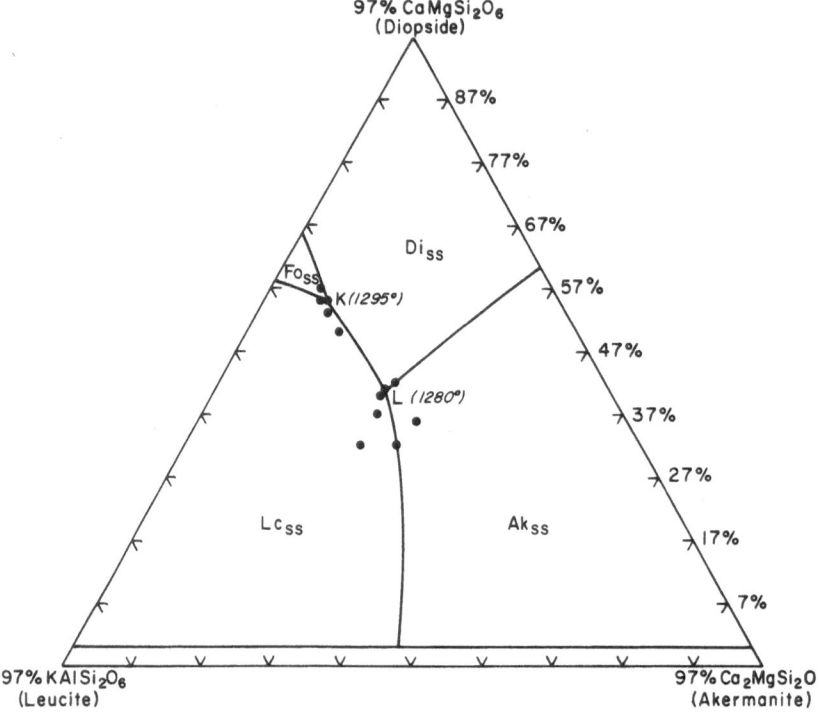

Fig. 10.4. The join diopside-akermanite-leucite with 3% forsterite. (After Gupta 1972)

($Fo_3Di_{55}Ak_9Lc_{33}$) forsterite$_{ss}$ + diopside$_{ss}$ + leucite$_{ss}$ + liquid are in equilibrium at 1295° ± 3 °C, and at L ($Fo_3Di_{41}Ak_{24}Lc_{32}$) diopside$_{ss}$ + akermanite$_{ss}$ + leucite$_{ss}$ + liquid coexist in equilibrium at 1280 °C, below which there is a liquid phase, which does not freeze until 1255 °C. As the compositions of the crystalline phases lie outside the join and there is a liquid phase below 1274 °C, the system forsterite-diopside-akermanite-leucite is not quaternary. Optical and X-ray study, however, suggests that although they are chemically nonstoichiometric, all the phases in this join contain only small amounts of solid solution and the behavior of the join thus approximates that of a quaternary system and points K and L behave as piercing points. From these two points liquid continues to crystallize along two different paths and at 1274 °C a five-phase assemblage of forsterite$_{ss}$, diopside$_{ss}$, akermanite$_{ss}$, leucite$_{ss}$, and liquid is reached. Below 1274 °C crystallization continues as the liquid moves toward the lowest melting point. Study of the join suggests that the lowest melting point of the tetrahedron, forsterite-diopside-akermanite-leucite, lies near the diopside-akermanite-leucite-plane. The final composition of the liquid thus lies near this plane.

10.6 Paragenesis

10.6.1 Melilite Leucitite. The join diopside-akermanite-leucite (Fig. 10.1) shows that a melilite leucitite, corresponding to the piercing point B, can be derived from either a melilite italite (aB), leucitite (bB), or a melilite (cB). Natural melilites contain a considerable amount of sodium in solid solution (Schairer and Yoder 1964a) but are poor in K_2O. The melilitites referred to here have compositions lying within the tetrahedron forsterite-diopside-akermanite-leucite.

At Capo di Bove in Italy, Washington (1906) noted the sporadic occurrence of leucitite grading to melilite leucitite (cecilite). Results of the system diopside-akermanite-leucite indicate that melilite leucitite may be derived from a leucitite magma.

10.6.2 Olivine Leucitite. Figure 10.2 shows that olivine leucitite (piercing point c) can be derived from either potassium-rich olivine pyroxenite (gc), or olivine italite (lc), or leucitite (ic). Olivine pyroxenite ordinarily contains very small amounts of K_2O. The olivine pyroxenites referred to here have compositions lying in the tetrahedron forsterite-diopside-akermanite-leucite. In the Bufumbira region of Uganda, Holmes (1937) described a lava flow of olivine leucitite (ugandite). At Katunga in Uganda, he noted the close association of rocks such as leucitite and olivine leucitite, and also reported the complete gradation of these rocks to pyroxene-rich leucitite, which in turn grades into pyroxenite. This field evidence and the results on the system forsterite-diopside-leucite suggest that an olivine leucitite magma can be a derivative of a potassium-rich olivine pyroxenite liquid. Generation of such a liquid is possible by the partial melting of phlogopite-bearing peridotite (Yagi and Matsumoto 1966).

10.6.3 Katungite. Phase equilibria study of the system forsterite-akermanite-leucite (Fig. 10.3) suggests that a katungite magma can be derived from a melilite italite (kG) or an olivine italite (lG). Field evidence of a massive lava flow of katungite was reported by Holmes (1937) from Toro-Ankole, Uganda, where it occurs in association with alnoite. On the basis of this experimental study it appears possible that a katungite magma may have been derived from an alnoite, which itself may have originated from a monticellite alnoite.

10.6.4 Olivine Melilitite. Reference to Ferguson and Merwin's diagram (1919) of forsterite-diopside-akermanite system suggests that olivine melilitite may be produced from peridotite, melilitite, or alnoite magmas.

10.7 Course of Crystallization of Liquid in the System Forsterite-Diopside-Akermanite-Leucite

The system is pseudoquaternary, being a part of the quinary system $K_2O-Al_2O_3-CaO-MgO-SiO_2$. The compositions of all crystalline phases lie outside the tetrahedron. However, the range of solid solution in these phases is small, and the system can thus be treated almost as quaternary. The complete course of crystallization of liquid in this system is summarized by a flow sheet diagram, shown in Fig. 10.5, in

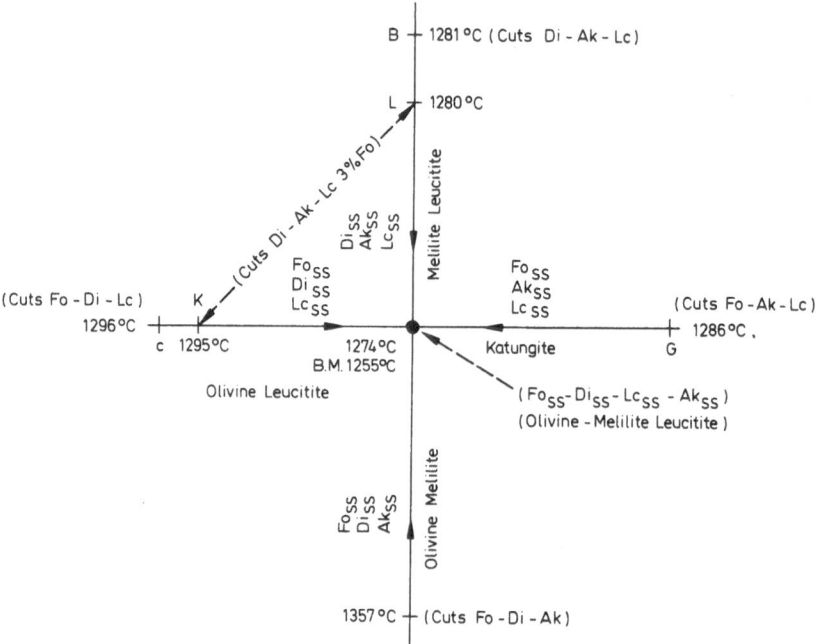

Fig. 10.5. The flow sheet and rock-nomenclature diagram of the system forsterite-diopside-akermanite-leucite

which a rock nomenclature is also given. It shows that an olivine-melilite-leucitite can be produced from a melilite leucitite, an olivine leucitite, a katungite or an olivine melilitite. Reference to Fig. 10.3 suggests that a katungite (point G) can be produced from an alnoite (point H). Schairer and Yoder (1964a), in their study on the system nepheline-forsterite-larnite-silica, showed that a parental liquid for the sodium-rich undersaturated part may be represented by either an olivine melilitite or an olivine nephelinite. Olivine-melilite nephelinite is produced from liquids of these compositions. It appears that rocks called olivine melilitite can differentiate toward either an olivine-melilite nephelinite or an olivine-melilite leucitite. The Na_2O/K_2O ratio of the parental olivine melilitite probably controls the subsequent of crystallization. If the liquid is potassium-rich, leucite-bearing mafic and ultramafic rocks would be produced, and if sodium-rich, nepheline-bearing mafic and ultramafic rocks would result.

Chapter 11 The System
Diopside-Nepheline-Akermanite-Leucite

The relationship of the potassium-rich mafic and ultramafic rocks with monticellite and olivine-bearing alnoites is understood by the study of the system forsterite-diopside-akermanite-leucite, in which nepheline is not included, to avoid complexity. However, the absence of nepheline is a serious loss to the system. Compositions of many other simplified potassium-rich rocks can be represented by the system diopside-nepheline-akermanite-leucite (Fig. 11.1). In the system under consideration, forsterite appears as a phase because of a reaction relationship between diopside and nepheline (Bowen 1922a; Schairer et al. 1962). Study of the system is thus important to understand the genesis of many other mafic and ultramafic volcanic rock types such as olivine-melilite-nepheline leucitite, nepheline-bearing katungite, and melilite-nepheline leucitite. Gupta et al. (1973b) studied this system and found that all phases appearing in this system are solid solutions. They established that nepheline incorporates variable amounts of kalsilite and excess silica; melilite contains soda-melilite and gehlenite; forsterite incorporates monticellite; diopside contains Ca-Tschermak's molecule and leucite incorporates nepheline in solid solution.

The system is bounded by four limiting joins: (1) diopside-nepheline-akermanite, (2) diopside-akermanite-leucite, (3) diopside-nepheline-leucite, and (4) nepheline-akermanite-leucite.

11.1 The Join Diopside-Nepheline-Akermanite

This join (Schairer and Yoder 1964; Onuma and Yagi 1967) is pseudoternary (Fig. 11.2) with two piercing points: one is located at $Di_{55}Ak_6Ne_{39}$ and $1212° ± 3 °C$ (G), where forsterite$_{ss}$, melilite; diopside$_{ss}$, and liquid are in equilibrium, and the other at $Di_{38}Ak_3Ne_{59}$ and $1169 °C$ (H), where forsterite$_{ss}$ + nepheline$_{ss}$ + melilite + liquid coexist. The assemblage forsterite$_{ss}$, diopside$_{ss}$, nepheline$_{ss}$, melilite, and liquid are in equilibrium at $1135 °C$. Forsterite$_{ss}$ has a reaction

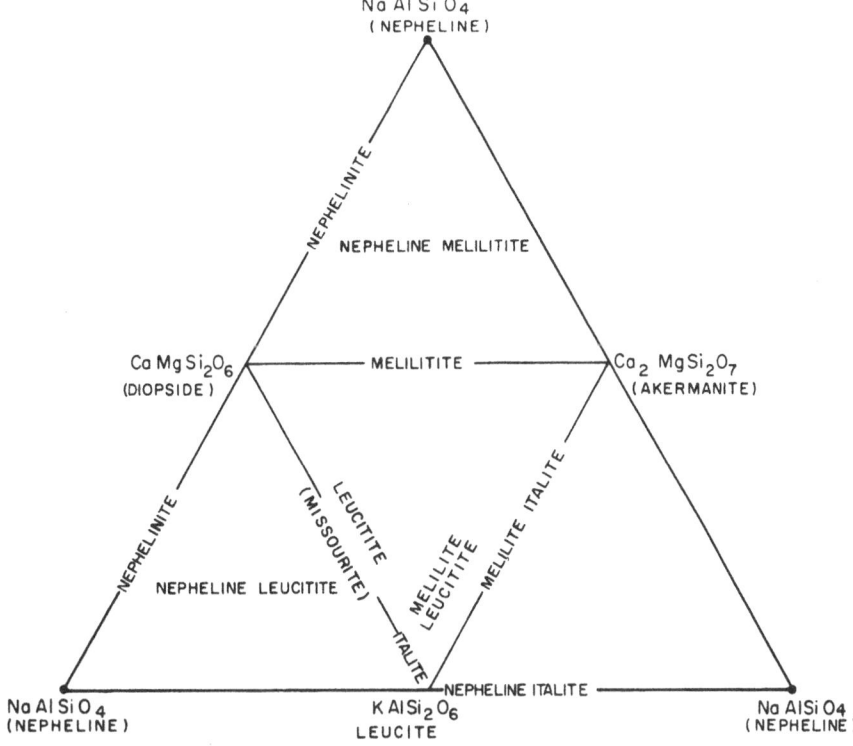

Fig. 11.1. Plot of compositions of leucite-bearing rocks in the system diopside-nepheline-akermanite-leucite

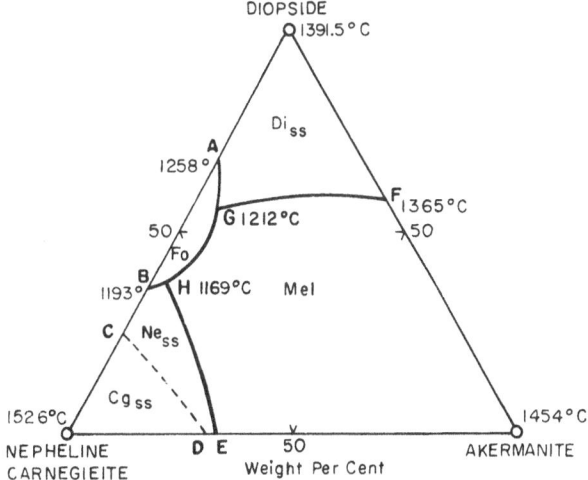

Fig. 11.2. Phase diagram of the join diopside-nepheline-akermanite at 1 atm. (After Onuma and Yagi 1967)

relationship with liquid. The join diopside-akermanite-leucite has already been discussed (Chap. 10.1). It has a piercing point at $Di_{39}Ak_{29}$-Lc_{32} and $1281° ± 3 °C$.

11.2 The Join Diopside-Nepheline-Leucite

The join shown in Fig. 11.3 (Gupta and Lidiak 1973) is pseudoternary with two four-phase points: one at $Di_{60}Ne_8Lc_{32}$ and $1275° ± 5 °C$ (A), where forsterite$_{ss}$, diopside$_{ss}$, leucite$_{ss}$, and liquid are in equilibrium, the second at $Di_{27.5}Ne_{29.5}Lc_{43}$ and $1194° ± 5 °C$ (B) where nepheline$_{ss}$, leucite$_{ss}$, and forsterite$_{ss}$ coexist with liquid. In this join, the assemblage forsterite$_{ss}$ + nepheline$_{ss}$ + diopside$_{ss}$ + leucite$_{ss}$ + liquid is reached between $1168°$ and $1100 °C$. Then forsterite$_{ss}$ reacts with liquid to be completely consumed. Near point (A) it disappears at $1135°$

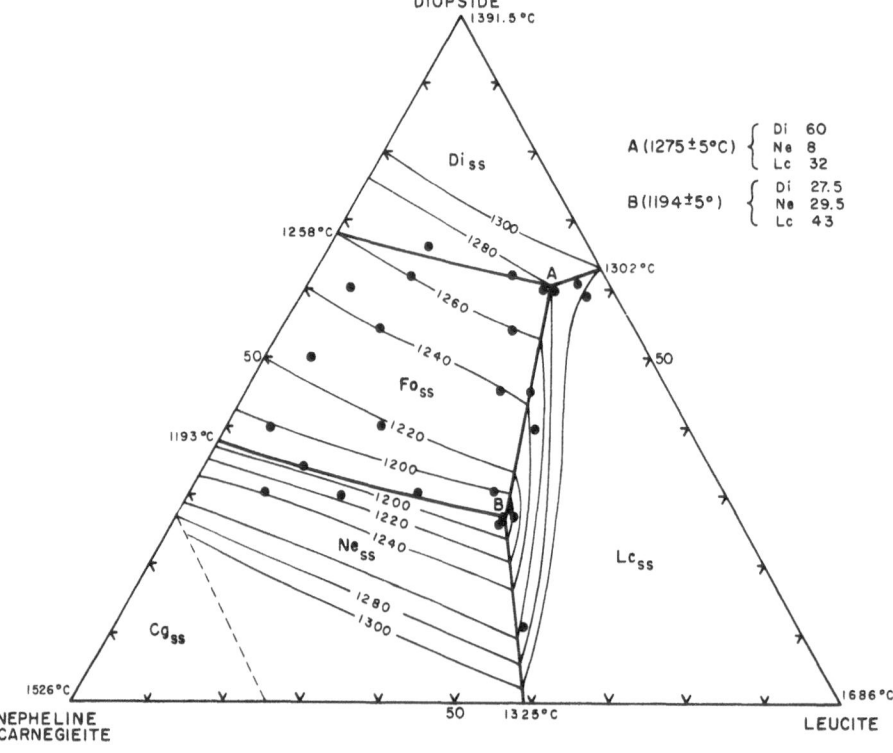

Fig. 11.3. Phase diagram of the join diopside-nepheline-leucite at 1 atm. (After Gupta and Lidiak 1973)

± 10 °C, whereas near point (B) it reacts at 1060° ± 10 °C. Near the join diopside-nepheline it disappears at 950° ± 10 °C. The final assemblage in this join consists of melilite + diopside$_{ss}$ + nepheline$_{ss}$ + leucite$_{ss}$. The course of crystallization of liquid in this join is shown by Fig. 11.4 and the rock nomenclature diagram corresponding to Fig. 11.4 is shown by Fig. 11.5.

11.3 The Join Nepheline-Akermanite-Leucite

This join was studied by Gupta et al. (1973b); the phase diagram of which is given in Fig. 11.6. At point S (Ne$_{34.5}$Ak$_{20.5}$Lc$_{45}$) and 1170° ± 3 °C, nepheline$_{ss}$, leucite$_{ss}$, melilite, and liquid are in equilibrium. Onuma and Yagi (1967) showed that in the join diopside-nepheline-akermanite, forsterite$_{ss}$ appears as the last solid phase. This phase also

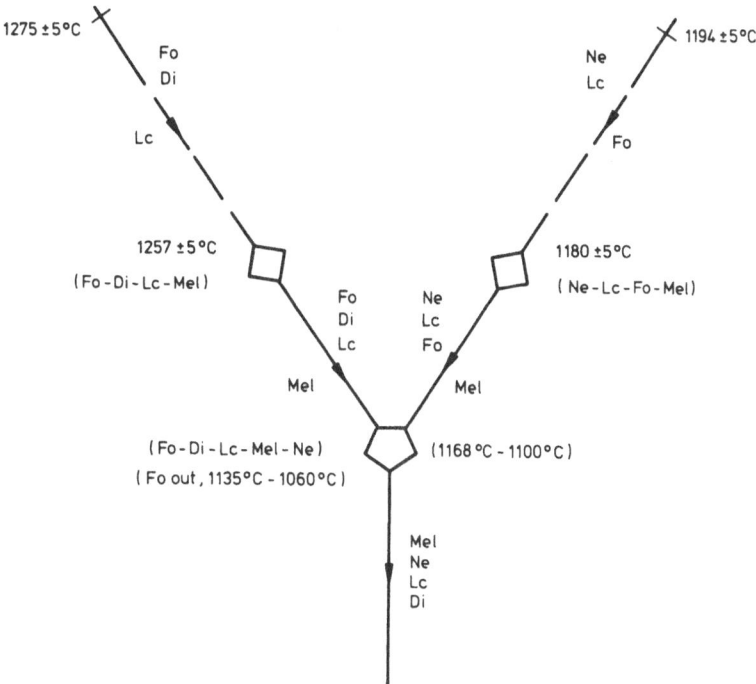

Fig. 11.4. Course of crystallization of liquid in the join diopside-nepheline-leucite. (After Gupta and Lidiak 1973)

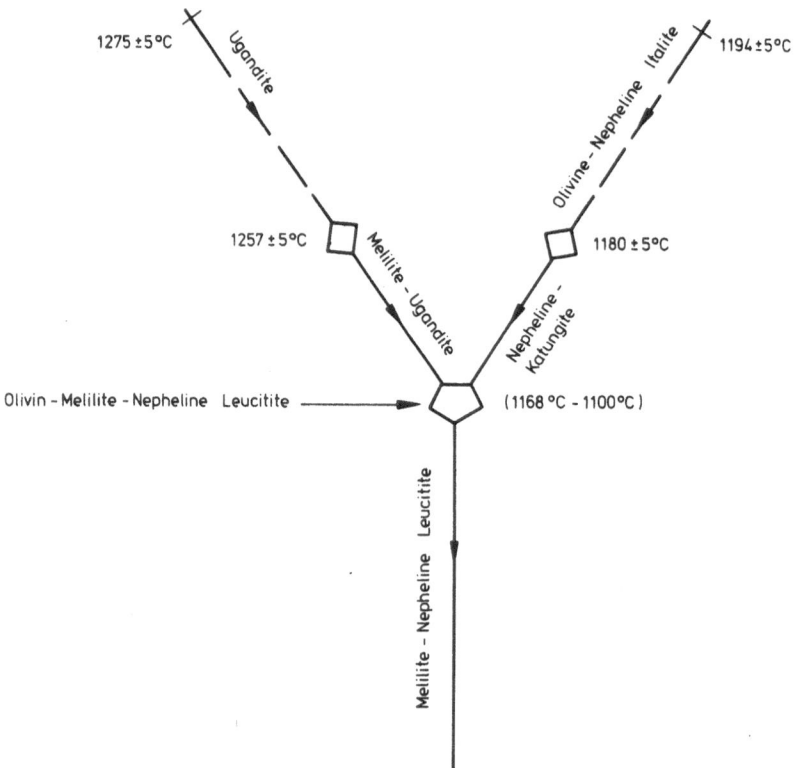

Fig. 11.5. Rock-nomenclature diagram, corresponding to Fig. 11.4. (After Gupta and Lidiak 1973)

appears as a subliquidus phase within the join nepheline-akermanite-leucite, but it reacts with the liquid and disappears at 1100° ± 5 °C. As melilite contains soda-melilite, and forsterite$_{ss}$ appears as a phase, the join is pseudoternary.

11.4 The Join $(Di_{38}Ak_3Ne_{59})_{100-x} - Lc_x$

Of the two piercing points in the system diopside-nepheline-akermanite (Onuma and Yagi 1967), point H (Fig. 11.2) has a relatively lower liquidus temperature. This point was therefore chosen as one of the end

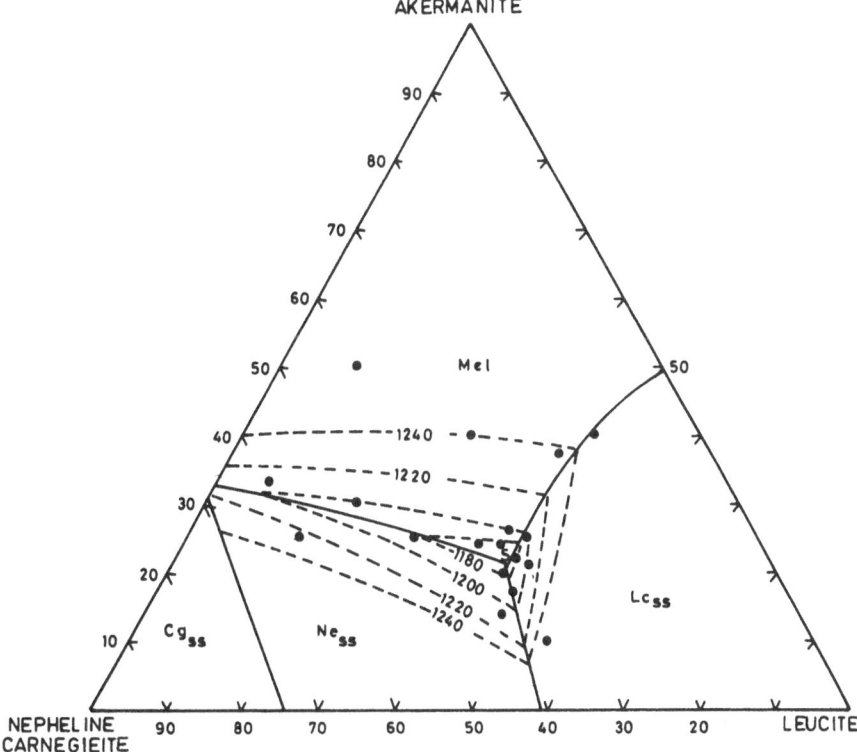

Fig. 11.6. Phase diagram of the join nepheline-akermanite-leucite at 1 atm. (After Gupta et al. 1973b)

members to which leucite was progressively added. The join was studied to determine the temperature of the six-phase assemblage (forsterite$_{ss}$ + diopside$_{ss}$ + akermanite$_{ss}$ + leucite$_{ss}$ + nepheline$_{ss}$ + liquid) in the system diopside-nepheline-akermanite-leucite, and the phase equilibrium diagram is presented in Fig. 11.7. The join cuts the primary phase volumes of forsterite$_{ss}$ and leucite$_{ss}$. The pseudoeutectic occurs at $(Di_{38}Ak_3Ne_{59})_{61}Lc_{39}$ and $1130° \pm 5 °C$, where forsterite$_{ss}$, leucite$_{ss}$, and liquid are in equilibrium. Figure 11.7 shows that at $1105° \pm 5 °C$, forsterite$_{ss}$ + diopside$_{ss}$ + melilite + leucite$_{ss}$ + nepheline$_{ss}$ coexist with liquid. Forsterite$_{ss}$ starts to react with liquid and is eliminated at $1075° \pm 10 °C$.

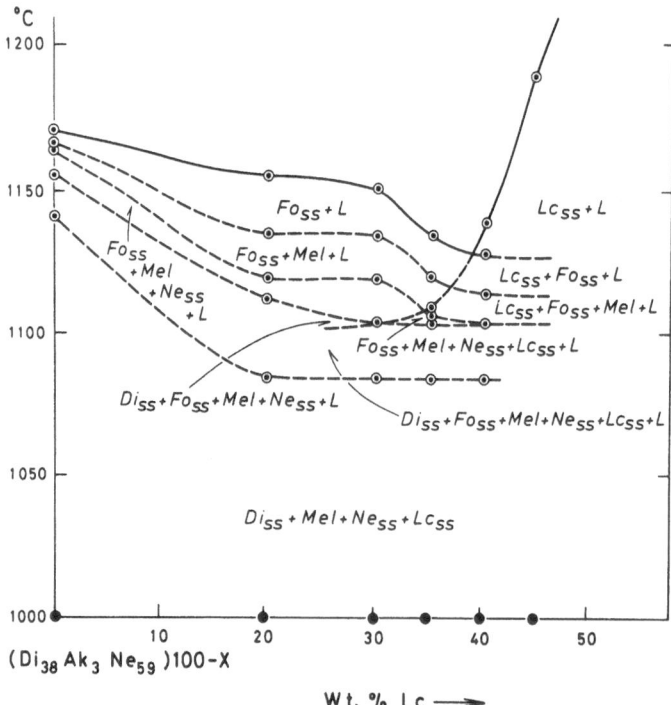

Fig. 11.7. Phase diagram of the join $(Di_{38}Ak_3Ne_{59})_{100-X}-Lc_X$ at 1 atm. (After Gupta et al. 1973)

11.5 Course of Crystallization of Liquid Within the System Diopside-Nepheline-Akermanite-Leucite

Experimental results on the join nepheline-akermanite-leucite, the pseudobinary join $(Di_{38}Ak_3Ne_{59})_{100-X}Lc_X$ and the other three bounding joins shows that the system diopside-nepheline-akermanite-leucite is pseudoquaternary. If, however, the presence of small amounts of alumina as Ca-Tschermak's molecule in diopside and as the gehlenite molecule in melilite is ignored, the system can be treated as quaternary join of the five component system, nepheline-kalsilite-CaO-MgO-SiO₂. A flowsheet diagram of the system is shown in Fig. 11.8, where only the five-phase univariant lines are shown. At A (Fig. 11.3) forsterite$_{ss}$, diopside$_{ss}$, leucite$_{ss}$, and liquid are in equilibrium at 1275° ± 5 °C. Gupta and Lidiak (1973) found that with further crystallization, the five-phase assemblage of forsterite$_{ss}$ + diopside$_{ss}$ + leucite$_{ss}$ + melilite

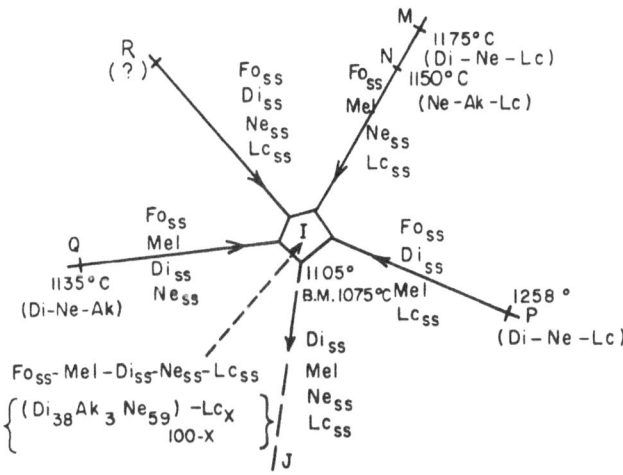

Fig. 11.8. Flowsheet diagram of the system diopside-nepheline-akermanite-leucite. (After Gupta et al. 1973)

+ liquid is reached at $1258° \pm 5$ °C. The above assemblage is represented by point P of Fig. 11.8. At B (Fig. 11.3) forsterite$_{ss}$, nepheline$_{ss}$, and leucite$_{ss}$ coexist with liquid. As the temperature drops melilite appears at 1175 °C. The assemblage of forsterite$_{ss}$ + nepheline$_{ss}$ + leucite$_{ss}$ + melilite + liquid is shown by point M (Fig. 11.8). In the join nepheline-akermanite-leucite (Fig. 11.6), nepheline$_{ss}$, melilite, leucite$_{ss}$, and liquid are in equilibrium at point S at 1170 °C. As crystallization continues the same assemblage of point M is reached at 1150 °C (N, Fig. 11.8). Onuma and Yagi (1967) found that in the join diopside-nepheline-akermanite, the five-phase assemblage of diopside$_{ss}$ + nepheline$_{ss}$ + forsterite$_{ss}$ + melilite + liquid is reached at 1135 °C, which is represented by point Q (Fig. 11.8). The six-phase invariant assemblage at I implies the possible existence of a five-phase piercing point within the system nepheline-kalsilite-CaO-MgO-SiO$_2$, where diopside$_{ss}$, nepheline$_{ss}$, leucite$_{ss}$, forsterite$_{ss}$, and liquid coexist. The join which contains this piercing point is unknown. The temperature of the six-phase assemblage at I is obtained from the pseudobinary system. After the disappearance of forsterite$_{ss}$, liquid moves from this point toward an invariant point of unknown composition. The disappearance of forsterite$_{ss}$ takes place over a temperature range of 30 °C (Fig. 11.8). The system diopside-nepheline-akermanite-leucite thus cannot be treated as quinary in a strict sense. A rock nomenclature diagram corresponding to Fig. 11.8 is presented in Fig. 11.9.

11.6 Paragenesis

Paragenetic relationship based on the field occurrence of potassium-rich mafic and ultramafic lavas are not clear. However, on the basis of the present phase equilibrium studies, some suggestions on the genetic relationship between various rock types can be made.

Holmes (1950) described the close association of rock types such as leucite katungite (1), olivine-melilite leucitite (2), melilite leucitite (3), leucite ankaratrite (4), and kalsilite-rich ultramafic rocks (5) from the Toro-Ankole field of East Africa. Rock types (1) and (2) may be represented by points M and P respectively of Figs. 11.8 and 11.9, and melilite leucitite corresponds to point B (Gupta 1972; Fig. 10.1). The join diopside-nepheline-leucite (Fig. 11.3) of Gupta and Lidiak (1973) shows that the assemblage corresponding to nepheline-leucite katungite (Fig. 11.5) can be obtained from an olivine-nepheline italite. Holmes and Harwood (1937) found katungite in the region between Bushwaga and Goma of the Birunga area of Uganda. The assemblage of P (olivine-melilite leucitite) is obtained by crystallization of a liquid from A (Fig. 11.3). Rock types corresponding to olivine leucitite sometimes containing melilite have been reported from more than a dozen localities by Holmes and Harwood (1937; pp. 81–82). Important localities include the Lutale flow, south-southwest of Lutale ridge and the Mikeno and Kisi areas (north-east corner of Lake Kivu). Olivine leucitite lavas

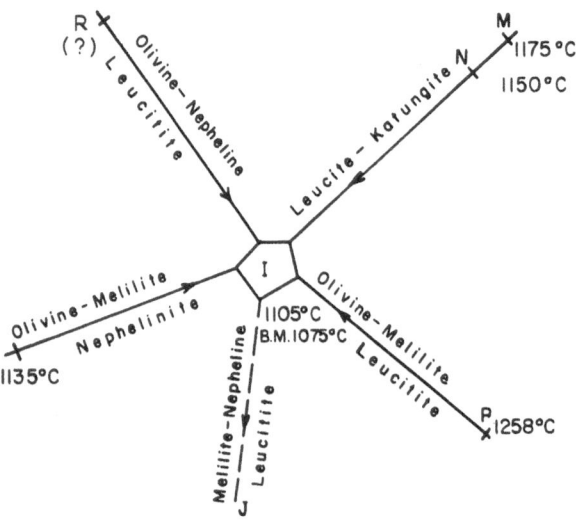

Fig. 11.9. Rock nomenclature diagram corresponding to Fig. 11.8

with high concentration of olivine (ugandite) have been reported by Holmes and Harwood (1937; p. 60–72) from various areas. Other localities include Muganza, Murehi, Musonga, and Katarara. The lava flows, corresponding to point R (olivine nepheline leucitite), have been described by the same authors (p. 79) from the following areas of the same volcanic field of Uganda: Mabungo, Duzakara, Bunagana, and Hyamichunchu. They also described lava flows of melilite-nepheline leucitite (IJ, Figs. 11.8 and 11.9) from Goma (p. 83). Melilite-nepheline leucitite (Fig. 11.9) can be obtained from olivine-melilite-nepheline leucitite (I), which has been reported from the Fort Portal area of Toro-Ankole by Holmes and Harwood (1932; p. 379), which belongs to the same petrographic province of the Birunga volcanic field. Melilite-nepheline leucitite has also been reported from the Villa Senni area of Italy by Washington (1906); the rarity of this rock type compared to other rock types represented by the flowsheet diagram (Fig. 11.8) is probably related to incomplete crystallization of magma.

11.7 Melilite-Plagioclase Incompatibility Problem in Leucite-Bearing Lavas

In the lavas of alkalic suites there is an incompatible relationship between melilite and plagioclase (Yoder and Schairer 1969; Yoder 1973). From volcanic centers of the same petrographic province, either melilite-bearing lavas are extruded or pyroclastics are ejected. Petrologic study of the melilite-bearing potassic rocks and leucite-bearing basanites from the Eifel area by the present authors support the conclusions of Yoder and Schairer (1969). From his study of natural rocks from various areas Yoder (1973) established that in alkalic lavas melilite and plagioclase do not coexist. Coexistence of these two minerals is, however, known in metamorphic rocks (Yoder 1973).

Gupta and Lidiak (1973) studied the system diopside-nepheline-leucite (Chap. 11.2) and found that the final assemblage consisted of diopside$_{ss}$, nepheline$_{ss}$, leucite$_{ss}$, and melilite. A mixture ($Di_{27}Ne_{29}$-Lc_{44}), containing the same assemblage, was chosen as one of the end members of a pseudobinary system to which anorthite was added to study the incompatibility between the mineral pair melilite and plagioclase. The phase diagram of the system $(Di_{27}Ne_{29}Lc_{44})_{100-X}$-$An_X$ is given in Fig. 11.10, which shows the final assemblage in the system to be diopside$_{ss}$ + nepheline$_{ss}$ + leucite$_{ss}$ + melilite + plagioclase + corun-

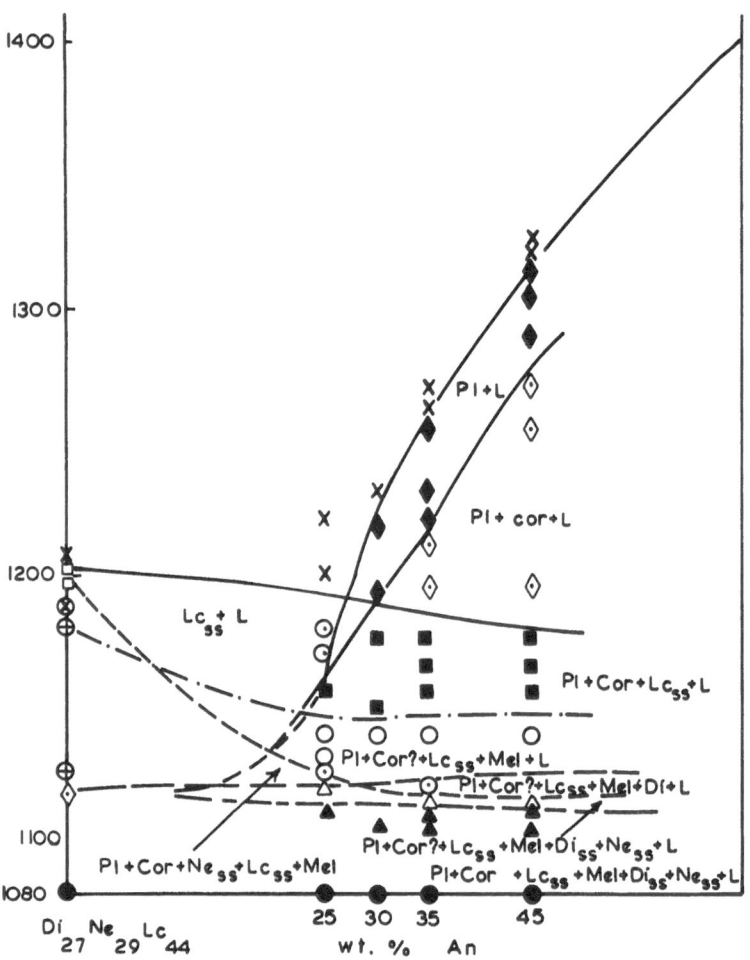

Fig. 11.10. Phase diagram of the join $(Di_{27}Ne_{29}Lc_{44})_{100-X}An_X$ at 1 atm

dum. Composition of melilite, as determined from the mixture Di_{27}-$Ne_{29}Lc_{44}$ at 1100 °C by the method of Hamilton and MacKenzie (1960), is $Ak_{64}Sm_{26}Geh_{10}$. The microprobe analysis of plagioclase from a mixture of bulk composition $(Di_{27}Ne_{29}Lc_{44})_{75}An_{25}$ gave $An_{92}Ab_6Or_2$. In leucite-bearing rocks, plagioclase is found to be calcium-rich (Shand 1943; Savelli 1967), as found in this synthetic system. Schairer et al. (1965) plotted the compositions of natural melilites in a composition triangle, where the three end members were $CaNaAlSi_2$-O_7, $Ca_2Al_2SiO_7$, and $Ca_2MgSi_2O_7$. When the composition of the melilite in the mixture of composition $(Di_{27}Ne_{29}Lc_{44})_{75}An_{25}$ is plotted in this triangle, it falls in the field of natural melilite. Yoder and Schairer (1969) studied the system akermanite-albite-anorthite at 1 atm and found that plagioclase and melilite are compatible with liquid over much of the field of plagioclase, with the exception of compositions on the join akermanite-albite. In a diagram of ln a_{SiO_2} vs T °C, Carmichael et al. (1970) plotted various silication reactions (Chap. 6.2, Fig. 6.2), which indicate that silication of forsterite + akermanite to produce diopside takes place much before the silication of nepheline to albite. The phase diagram of the join albite$_{50}$akermanite$_{50}$–anorthite$_{50}$akermanite$_{50}$, studied at 1 atm by Yoder and Schairer (1969; Fig. 11.11) shows the melilite-plagioclase coexistence over a broad temperature-composition range. These authors also studied a natural

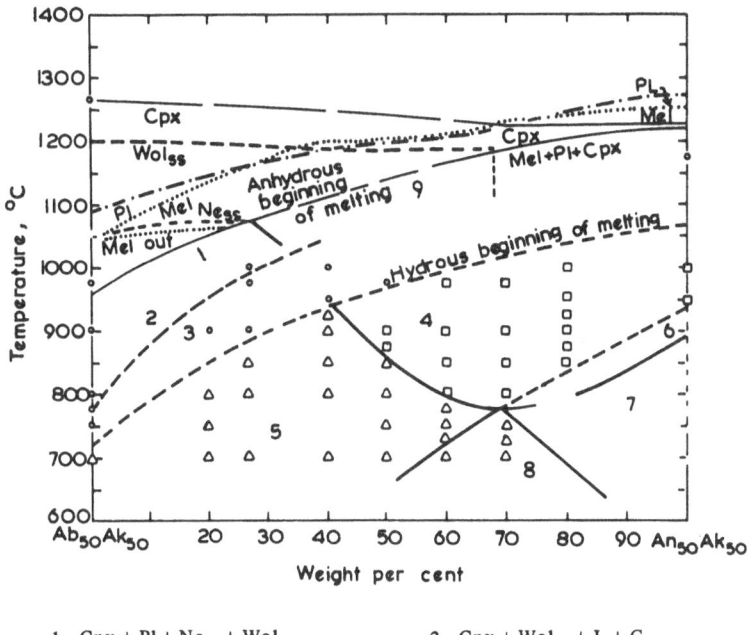

1. Cpx + Pl + Ne$_{ss}$ + Wol$_{ss}$ 2. Cpx + Wol$_{ss}$ + L + G

3. Cpx + Wol$_{ss}$ + Pl + L + G   4. Cpx + Mel + PL + Wol$_{ss}$ + G

5. Cpx + Pl + Ne + Wol$_{ss}$ + G   6. Cpx + Mel + Pl + Gr + Wol$_{ss}$ + G

7. Cpx + Gr + Pl + Wol$_{ss}$ + G   8. Cpx + Pl + Wol$_{ss}$ + Ne$_{ss}$ + Gr + G

9. Mel + Pl + Cpx + Wol$_{ss}$

Fig. 11.11. Phase diagram of the join albite$_{50}$ akermanite$_{50}$–anorthite$_{50}$ akermanite$_{50}$ at 1 atm. (After Yoder and Schairer 1969). See also List of Abbreviations

melilite (akermanite:sodamelilite is 2:1) and natural plagioclase (close to anorthite$_{50}$) mixed in equal proportions, and noted that above 850 °C melilite and plagioclase coexisted with diopside$_{ss}$, wollastonite$_{ss}$, and liquid. However, under 2 kb P_{H_2O}, the assemblage consisted of diopside$_{ss}$ + plagioclase + nepheline$_{ss}$ + wollastonite$_{ss}$ at 700 °C, 800 °C, and 900 °C.

A mixture of composition $(Di_{27}Ne_{29}Lc_{44})_{75}An_{25}$ was studied under water pressures by Gupta (unpublished) to see if the presence of water in the system has any effect on the melilite-plagioclase coexistence. The study showed that below 2.5 kb at temperatures of 700° and 800 °C the assemblage consisted of diopside$_{ss}$ + nepheline$_{ss}$ + melilite + anorthite$_{ss}$ + liquid + vapor. Above 2.5 kb phlogopite appeared as an important phase and leucite$_{ss}$ disappeared, but melilite coexisted with anorthite$_{ss}$, phlogopite$_{ss}$, nepheline$_{ss}$, liquid and vapor. Above 2.8 kb melilite disappeared and grossularite$_{ss}$ appeared. The equilibrium assemblage at 700° and 750 °C and 5 kb consists of large quantities of grossularite$_{ss}$, phlogopite$_{ss}$, small amounts of diopside$_{ss}$ and rare anorthite$_{ss}$. Yoder (1969) studied a mixture of composition $Ak_{50}An_{50}$ at different temperatures and water vapor pressures (Fig. 11.12) and found that at the low temperature region akermanite and anorthite reacted to produce grossularite$_{ss}$ and diopside$_{ss}$ and at high temperatures and moderate pressures the assemblage consisted of melilite + anorthite + diopside$_{ss}$ + wollastonite$_{ss}$. In his starting materials there was neither nepheline$_{ss}$ nor leucite$_{ss}$ as in the present study. The study of the mixture $(Di_{27}Ne_{29}Lc_{44})_{75}An_{25}$ at different temperatures and water pressures shows that calcium-rich plagioclase does not coexist with melilite at higher P_{H_2O}, as the reaction diopside and nepheline to produce melilite in the low temperature region of the system diopside-nepheline (Schairer et al. 1962) ceases to exist at high water pressures. However, this mineral pair coexist at low water pressures under volcanic and subvolcanic conditions.

Schairer and Yoder (1970) studied the system CaO–MgO–Al$_2$O$_3$–SiO$_2$, and found that in the volume enclosed by anorthite-akermanite-diopside-forsterite-spinel, akermanite coexists with spinel, anorthite and diopside in the absence of forsterite. Such a calcium-rich assemblage is found only in metamorphic rocks, whereas the assemblage forsterite-akermanite-diopside-spinel has representatives among the igneous rocks. They indicated that olivine may be a deciding factor in the melilite-plagioclase incompatibility in the lavas. Thus in the absence of olivine, melilite, and plagioclase may coexist in some hybrid

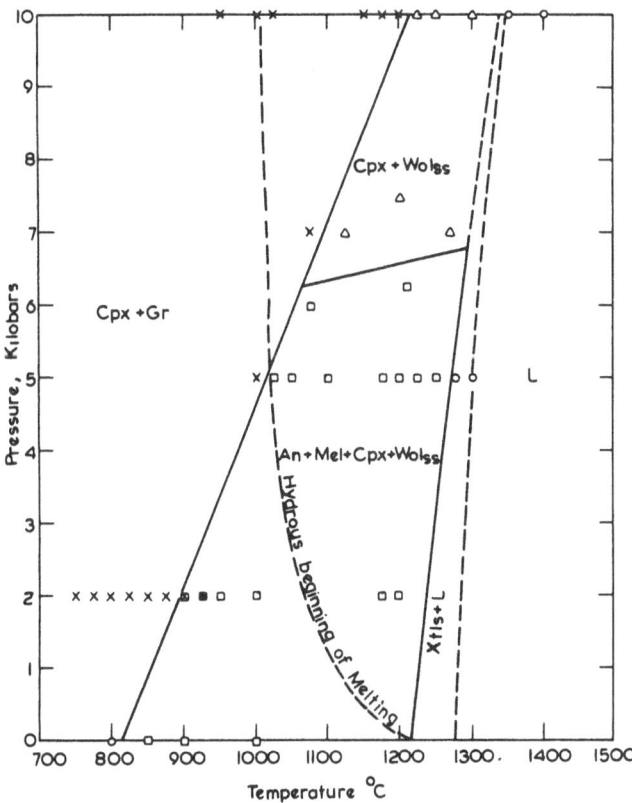

Fig. 11.12. P–T diagram for the bulk composition anorthite$_{50}$ akermanite$_{50}$ by weight. *Xtls* various assemblages. (After Yoder 1969). See also List of Abbreviations

contact zone, and in presence of water, even in the absence of olivine there is a limit to the anorthite content of plagioclase that can coexist with melilite (H.S. Yoder 1977, personal communication).

Chapter 12   Solubility of $KFe^{3+}Si_2O_6$ in Leucite

Published chemical analyses of leucites (Deer et al. 1963) show that their departure from stoichiometric composition is not very significant. Most studies on leucite$_{ss}$ are related to the substitution of potassium by sodium, which was considered to be very limited by Bowen and Ellestad (1937). Experimental studies by Fudali (1963) established that such a substitution is more extensive in the system $KAlSi_2O_6$–$NaAlSi_2O_6$. Incorporation of the compound $K_2MgSi_5O_{12}$ by leucite has been studied by Schairer (1948), who established a complete series of solid solution through the substitution of $MgSi \rightleftharpoons AlAl$ between $K_2Al_2Si_4O_{12}$ and $K_2MgSi_5O_{12}$. However, no experimental data are available on the extent of solubility of iron in leucite. The reported chemical analyses of leucite (Faust 1963; Deer et al. 1963) show a prevalence of Fe^{3+} over Fe^{2+}, which can be related to substitution of Al^{3+} by Fe^{3+} in the tetrahedral sites. Leucites from potassic lavas of the Leucite Hills contain more than 2% Fe_2O_3 (Carmichael 1967) and there is a relative excess of Si and K over Al. Leucites from other areas show much lower contents of iron (Faust 1963) in spite of the similar bulk compositions of the host rocks. Stability of the $KFeSi_2O_6$ (FeLc) compound and its crystal structure was studied by Faust (1963). The extent of solid solution of $KFe^{3+}Si_2O_6$ in leucite at 1 atm and at pressures of 1 and 2 kb will be discussed in the following pages. The study was made under partial pressure of H_2O_2 in order to ensure highly oxidizing conditions.

Three synthetic mixtures of compositions $Lc_{90}(FeLc)_{10}$, Lc_{94}-$(FeLc)_6$, and $Lc_{97}(FeLc)_3$ were prepared. The mixtures were homogenized by sintering at 1350 °C for half an hour followed by repeated crushing and heating three times and then crystallized at 1000 °C for at least three days. Both homogeneous mixtures and sintered materials were used as starting materials in the experiments. The results were identical in both cases.

The results of the investigation are summarized in Table 12.1, which includes the results of the more significant runs. It is found that leucite accepts only a limited amount of Fe^{3+} (less than 6 wt. % $KFeSi_2O_6$) in solid solution at atmospheric pressure. In the compositional range between $Lc_{94}(FeLc)_6$ and $Lc_{90}(FeLc)_{10}$, leucite$_{ss}$ coexists with hematite and high sanidine, which can be explained by the following reaction:

$$2\ KAlSi_2O_6 + 2\ KFeSi_2O_6 \rightarrow Fe_2O_3 + 2\ KAlSi_3O_8$$
(leucite)   (Fe-Leucite)   (hematite)   (sanidine)

$$+ K_2Si_2O_5$$
(K-disilicate)

$K_2Si_2O_5$ probably is incorporated by leucite in limited amount as solid solution and thus does not appear as an independent phase.

The solubility of $KFeSi_2O_6$ in leucite, however, increases slightly with P_{H_2O} in comparison to the runs at 1 atm. Electron microprobe analyses of leucite from run 14 (Table 12.1) show that it contains 2.49 wt.% of Fe_2O_3 which corresponds to 7.71 wt.% $KFeSi_2O_6$. The presence of hematite suggests that the saturation level of Fe^{3+} under that condition was exceeded.

Contrary to leucite, nepheline can incorporate more Fe^{3+} in its structure. In their study on the join nepheline-"iron nepheline" ($NaFeSiO_4$), Onuma et al. (1972) found that nepheline can contain up to 30 mol% iron nepheline at 925 °C under atmospheric pressure. They established a gradual increase in the refractive indices and unit cell dimensions of nepheline$_{ss}$ with increasing $NaFeSiO_4$ contents. Fe_2O_3 is also soluble in microcline in the form of "iron orthoclase" ($KFeSi_3O_8$), but its solubility is much lower (Rosenqvist 1951).

Refractive indices of leucite$_{ss}$ were also measured and it was found that although the crystal structure of the phase was tetragonal, optically it was nearly isotropic with a birefringence of 0.001. The refractive indices of all leucite$_{ss}$ grains were found to vary within the range of 1.507 ± 0.003. Possibly because of limited substitution of Fe^{3+} in leucite$_{ss}$, variation in the values of indices of refraction was also insignificant.

The cell parameters of leucite$_{ss}$, crystallized under different temperatures and pressures, were determined accurately by X-ray. Table 12.1 indicates that all leucite crystals, formed under pressure, are slightly stretched along the c-axis in comparison with those crystallized under atmospheric pressure. The values of "a" of the leucites formed at 1 and 2 kb P_{H_2O} are lower in contrasts to those of the leucites heated at atmosphere, and natural leucites (Table 12.1). If the cell parameters

Table 12.1. Experimental results on three different mixtures of compositions, $Lc_{97}(FeLc)_3$ (1), $Lc_{94}(FeLc)_6$ (2), and $Lc_{90}(FeLc)_{10}$ (3) at 1 atm, 1 and 2 kb $P_{H_2O_2}$

No.	Composition	Pressure	Time (h)	Temperature (°C)	Phases	Cell parameters of leucite$_{ss}$[a]		
						a (Å)	c (Å)	V (Å)³
1	(1)	1 atm	19	1,060	Leucite$_{ss}$	13.104	13.759	2,362.8
2	(1)	1 atm	18	1,160	Leucite$_{ss}$	13.089	13.740	2,354.0
3	(1)	1 atm	5	1,210	Leucite$_{ss}$	13.091	13.749	2,356.2
4	(1)	1 atm	2	1,260	Leucite$_{ss}$	13.100	13.728	2,356.0
5	(1)	2 kb	98	650	Leucite$_{ss}$	13.066	13.778	2,352.2
6	(2)	1 atm	19	1,060	Leucite$_{ss}$	13.139	13.764	2,372.5
7	(2)	2 kb	48	650	Leucite$_{ss}$	13.086	13.780	2,359.8
8	(3)	1 atm	2	1,260	Leucite$_{ss}$ + sanidiness + hematite	13.105	13.744	2,360.2
9	(3)	1 atm	5	1,210	Leucite$_{ss}$ + sanidiness + hematite	13.107	13.757	2,363.5
10	(3)	1 atm	18	1,160	Leucite$_{ss}$ + sanidiness + hematite	13.101	13.763	2,362.4
11	(3)	1 atm	40	1,030	Leucite$_{ss}$ + sanidiness + hematite	13.098	13.751	2,358.8
12	(3)	1 kb	43	815	Leucite$_{ss}$ + sanidiness + hematite	13.072	13.823	2,362.2
13	(3)	1 kb	41	620	Leucite$_{ss}$ + sanidiness + hematite	13.077	13.782	2,356.8

14	(3)	2 kb	48	650	Leucite$_{ss}$ + sanidine$_{ss}$ + hematite	13.070	13.792	2,356.2
15	(Fe–Lc)[b]	1 atm	72	1,000	"iron leucite"	12.839	14.139	2,330.785

[a] Data of cell parameters are determined by R. Farinata, L. Loreto, and R. Trigilla of the University of Rome (December 1974, personal communication)

[b] Data of Gupta (unpublished)

of leucites are compared with those of pure iron leucite (Table 12.1), an increase in the values of "c", and a decrease in "a" values are expected with iron enrichment. Therefore the cell parameter data of leucite$_{ss}$, crystallized under pressure, are in agreement with what is expected. However, leucite crystallized at different temperatures at atmospheric pressure has higher values of "a" and lower values of "c" (Table 12.1), when compared with those of leucite$_{ss}$, heated under 1 and 2 kb. The cell volume data of this phase crystallized under both conditions are similar (Table 12.1). Natural leucites have always lower values of "a", "c", and cell volume and they form a homogeneous group regardless of their chemical composition and different genetic environments.

Results of the investigation suggest that substitution of Fe^{3+} in synthetic leucites does not show any systematic variation in cell parameters.

Chapter 13 Survival of Leucite

13.1 Alteration of Leucite to Analcite

Leucite is usually found in the rocks of Tertiary or younger age. Absence of leucite in the older rocks may be attributed to its alteration to analcite. From their experimental and theoretical studies Gupta and Fyfe (1975) demonstrated that, given sufficient time, leucite will not survive diagenetic processes, as the leucite-bearing rocks will convert to their analcite-bearing analogues before any major metamorphism occurs.

A natural leucite from Roccamonfina, Italy, was used by them for such a study. The chemical composition of the leucite is given in Table 3.1. The leucite was ground to 100 mesh and was allowed to react with salt solutions. In one set of experiments 20 mg of sodium chloride with 20% water was placed in sealed gold capsules and the reaction was studied at 1 kb total pressure. In this case the salt solution was saturated.

In a second set of experiments, 0.1 g of leucite was mixed with 2 cm³ of synthetic sea water (a salt mix of major species prepared for biological studies). These solutions contain sodium and potassium in normal sea water concentration. The results of the experiments are summarized in Fig. 13.1. The amount of conversion was estimated from

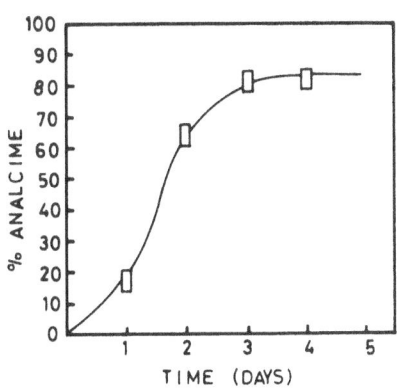

Fig. 13.1. Alteration of leucite to analcite. (After Gupta and Fyfe 1975)

X-ray diffraction patterns calibrated with known mixtures of analcite and leucite.

From the experimental results it is clear that the reaction is very fast and can be measured easily down to 150 °C. As the reaction is easily studied over nearly 100 °C range, it follows that the activation energy is small (Fyfe 1973). If we assume that for the early part of the reaction the rate equation is zero order, that is

$$\frac{dc}{dt} \simeq K$$

(note that the process is almost linear up to 80% conversion), then the activation energy is in the order of 8 kcal mol^{-1} and certainly less than 10 kcal. This is a very low activation energy and the reaction will be rather temperature-insensitive. If we consider that the time constants of diagenetic (burial) process may be in the order of 10^5–10^7 years, then even at 25 °C, given appropriate fluids, the reaction would go to completion.

The process being studied is:
$$KAlSi_2O_6 \text{ (solid)} + Na^+aq + H_2O \rightleftharpoons NaAlSi_2O_6 \cdot H_2O \text{ (solid)} + K^+aq$$
$$\Delta G^0_{298} = -1093 \text{ mol}^{-1}$$

(Robie and Waldbaum 1968). While there is considerable uncertainty in ΔG^0, this figure suggests that analcite is stable even when the concentration of potassium exceeds the concentration of sodium (K^+_{aq}/Na^+_{aq} = 6.4). This implies that even normal river water (K^+/Na^+ = 0.3) could cause the reaction.

Although no nucleation is involved in the reaction (but a 10% volume expansion occurs) a number of steps could be rate-determining, which might include:

a) diffusion rates of Na$^+$ or K$^+$ in the crystal,
b) dehydration rates of ions before entry,
c) diffusion rate of water into the crystal.

The overall ΔH (cf. ΔE) of processes like:

$$K^+_{solid} + Na^+_{aq} \rightleftharpoons Na^+_{solid} + K^+_{aq} \text{ are small (Latimer 1964)}.$$

Normal heats of hydration of solids are similar to the activation energy determined in their work. The fact that more dilute sea water appears to react faster than the saturated salt solutions may indicate that the water activity in diffusion is important.

Numerous authors note the facile leucite-analcite reaction (Deer et al. 1963; Bragg et al. 1965). The main conclusion of the work of Gupta and Fyfe (1975) is that if sufficient sodium-bearing water is available, then conversion will occur even at surface temperatures. The chances for leucite survival are small and in fact most reported occurrences of volcanic leucite are in rather recent rocks. It is therefore very possible that this process may occur at low temperatures when other primary igneous materials might show little alteration. This possible reaction must therefore be considered before analcite is considered as a primary igneous phase (Pearce 1970). Nakamura and Yoder (1974), who studied analcite "phenocrysts" in basalts, also concluded that they are most likely exchange products of original leucite.

13.2 Pseudoleucite and its Genesis

The name pseudoleucite is given to a certain aggregate, having the shape of trapezohedral crystals such as those of leucite and analcite. The chief constituents of the aggregates are K-feldspar and nepheline, occasionally with zeolite and white mica (Shand 1943; p. 448). The feldspar-nepheline intergrowth in pseudoleucites is often present in a zonal arrangement. For example Yagi (1954) found a narrow outer zone in pseudoleucite, consisting of orthoclase and nepheline, oriented randomly, while in the inner zone, the longer dimensions of nepheline, analcite, and K-feldspar were found to be oriented perpendicular to the crystal boundary. In the Serra de Caldas, Brazil, Hussak (1890) described pseudoleucite with a narrow mantle of orthoclase laths within which nepheline is also present. Most of the pseudoleucites in the Highwood Mountain areas are made up of a mixture of nearly pure potassic feldspar and a cloudy amorphous material. The composition of the amorphous material was found by Larsen and Buie (1938; p. 1840) to be equivalent to hydrated nepheline. They found that in many cases leucite and analcite have been converted completely to pseudoleucite, but sometimes this conversion has taken place only at the margins, leaving most of the crystals as fresh leucite and analcite. Bowen and Ellestad (1937) described this marginal alteration of leucite to pseudoleucite in the Nyamlagira area of Bufumbira volcanic field.

Pseudoleucites have been reported from various areas. Important localities include: Bearpaw Mountains and Highwood Mountains of Montana (USA); Spotted Fawn Creek, Yukon Territory (Canada);

Magnet Cove, Arkansas (USA), Loch Borolan Laccolith (Scotland),
Serra de Caldas (Brazil), Laacher See district (FRG); Tzu Shin Shan,
Shansi (N. China), and Tezhsarsk (Armenia). According to Zies and
Chayes (1960) pseudoleucites usually occur in tinguaites and rarely
in shonkinites and monchiquites. In the Highwood Mountains area
pseudoleucites are found in lava flows and shallow dikes in the fine-
grained groundmass (Larsen and Buie 1938). In the same area they
are also found in lavas and near surface intrusives. Shand (1910) has
reported plutonic pseudoleucites.

Chemical analyses of pseudoleucites determined by various investigators are shown in Table 13.1. If minor oxides are not considered
these compositions can be plotted in the nepheline-kalsilite-SiO_2
system of Schairer and Bowen (1935) shown in Fig. 13.2. Zies and
Chayes (1960) made micrometric analyses of pseudoleucites from
tinguaite dike in Bearpaw Mountains (Montana) and found that the
micrometric mode consisted of 29.8% nepheline and 66.2% K-feldspar. Yagi and Gupta (1977) studied the pseudoleucites from the
Tezhsarsk area of Armenia. These pseudoleucites are essentially composed of nepheline and K-feldspar with minor amounts of sericite,
analcite, kaolin, epidote, and biotite. Yagi and Gupta (1977) determined
the compositions of nepheline and feldspar by the X-ray powder diffraction technique. The composition of nepheline, determined by the
method of Hamilton and MacKenzie (1960) is $Ne_{87.9}Ks_{3.0}(SiO_2)_{9.1}$ and

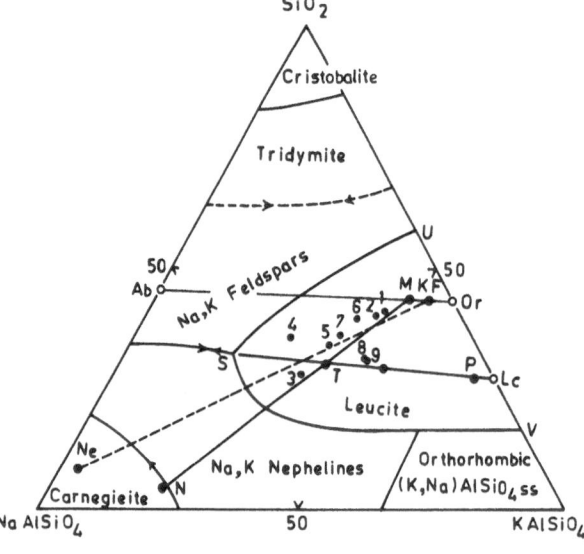

Fig. 13.2. Plot of the
compositions of pseu-
doleucite in the system
nepheline-kalsilite-SiO_2

Table 13.1. Analyses of pseudoleucites from various localities

	1	2	3	4	5	6	7	8	9
SiO_2	60.06	55.55	55.06	58.15	58.30	56.9	56.26	57.42	57.46
TiO_2	0.28	0.11	–	–	–	0.1	–	0.24	0.07
Al_2O_3	21.39	22.27	25.26	23.66	23.80	22.6	21.93	21.85	21.94
Fe_2O_3	0.29	1.66	–	1.59	–	0.6	0.67	1.70	0.66
FeO	N.D.	0.33	–	–	–	–	0.00	0.00	0.13
MnO	0.03	tr.	–	–	–	–	–	0.03	0.04
MgO	0.08	0.09	0.28	0.21	0.17	–	0.00	0.07	0.00
CaO	0.39	2.16	0.60	0.43	0.96	1.3	1.46	0.19	0.20
Na_2O	2.85	2.80	7.60	7.08	5.80	3.9	4.95	4.78	4.78
K_2O	12.33	10.37	10.34	8.49	10.94	10.8	10.63	13.40	13.76
H_2O^+	2.28	4.09	1.78	1.35	–	3.5	4.16	0.27	0.36
H_2O^-	0.07	0.99	–	–	–	–	–	0.03	0.00
Rest	–	–	–	–	–	–	–	0.28	0.50
Total	100.05	100.42	100.92	100.96	99.97	99.7	100.06	100.26	99.90

1. Pseudoleucite from Tezhsarsk, Armenia (Yagi and Gupta 1977)
2. Pseudoleucite from Tzu Shin Shan, China (Yagi 1954)
3. Pseudoleucite from Magnet Cove, Arkansas. Anal. C.W. Knight (Williams 1890; cited by Knight 1906)
4. Pseudoleucite from Spotted Fawn Creak, Yukon Territory Knight (1906)
5. Pseudoleucite from Vesuvius, Italy (Knight 1906)
6. Pseudoleucite from Highwood Mountains, Montana. Anal. F.A. Gonyer (Larsen and Buie 1938)
7. Pseudoleucite from Loch Borolan Laccolith, Scotland (Shand 1939)
8. Pseudoleucite from Bearpaw Mountains, Montana. Anal. C.O. Ingamel (Zies and Chayes 1960)
9. Pseudoleucite from Bearpaw Mountains, Montana (Fudali 1963)

that of alkali feldspar determined from the knowledge of its cell volume by the method of Orville (1967) is $Or_{92}Ab_8$.

Formation of pseudoleucite in the Vesuvius area was considered to be due to autopneumato-metamorphism (Rittmann 1933). Knight (1906) considered that after the entire magma had solidified, the sodium-rich leucite changed to pseudoleucite. Shand (1943, p. 448) described this reaction as follows: $2(K, Na) AlSi_2O_6 \rightleftharpoons KAlSi_3O_8 + NaAlSiO_4$. Yagi (1954) thought that the pseudoleucites in the Tzu Shin Shan area were also formed by the breakdown of a sodic leucite. Larsen and Buie (1938) thought that the mechanism described by Knight (1906) might have been responsible for the formation of pseudoleucite in the Highwood Mountain area but also suggested that some pseudoleucites might have been the result of the breakdown of potassium-rich analcite, and to support this view they showed that some of the analcites in this area contained up to 4.48% K_2O (21% leucite). The refractive indices of these analcites are consistently high (approximately 1.493). Fudali (1957) gave a similar view on the basis of his work on pseudoleucite in the Blue Mountain area of Ontario. However, in his later experimental work (Fudali 1963) he found that the maximum amount of solid solution in analcite was approximately 8%. However, he suggested that the limited amount of solid solution of leucite in analcite may be related to kinetics of reaction or low water pressure used in his study.

Bowen and Ellestad (1937) objected to the views of Knight (1906) and suggested on the basis of the study of the system nepheline-kalsilite-SiO_2 (Bowen and Schairer 1935) that leucite cannot incorporate more than 1%–1.5% Na_2O. Bowen and Ellestad (1937) suggested that if a leucite of composition P (Fig. 13.2) reacts with the liquid of composition S, it should give a solid product, corresponding closely with analyzed pseudoleucite as their compositions lie close to the line PS near point T. At T the ratio of the two reactants, liquid and leucite, would be equal to the ratio of PT/ST. Compositions of nepheline and K-feldspar produced by the reaction would be given by points N and M respectively. However, if from a liquid of composition P there is subtraction of some leucite as cooling continues the composition of the liquid may not reach point S but would reach either curve US or else curve VS. If the liquid reaches US, leucite would be transformed by reaction to K-feldspar only, and if it reaches curve VS it would react with the liquid to form nepheline. In turn Larsen and Buie (1938; p. 443) objected to the view of Bowen and Ellestad and favored the idea that pseudoleucite is formed by the inversion of potassium-rich analcite or sodium-rich leucite on the following basis:

1. Well-formed crystal pseudomorphs of leucite with sharp boundaries in rapidly cooled rocks.
2. The extent to which pseudoleucite formed did not depend on the rate of cooling, as many of the rocks with fresh leucite crystals were from coarse-grained stocks, and in dikes, pseudoleucite formed to about the same extent in the fine-grained chilled borders as in the coarse-grained centers.
3. In some of the lavas and near the contacts of some of the dikes where pseudoleucite is found the groundmass was submicroscopic, implying that cooling was rapid. Yet within 10 mm the replacement to pseudoleucite is complete. Larsen and Buie (1938) considered that the mechanism suggested by Bowen and Ellestad (1937) would involve slow reaction as it would require diffusion both through the leucite and interstitial liquid.

Fudali (1963) studied a selected part of the system nepheline-kalsilite-SiO_2 at 1 kb P_{H_2O} and elevated temperatures. He found that under this condition the primary phase field of leucite is considerably restricted and in the join $KAlSi_2O_6$–$NaAlSi_2O_6$ (Fig. 13.3) at 800 °C

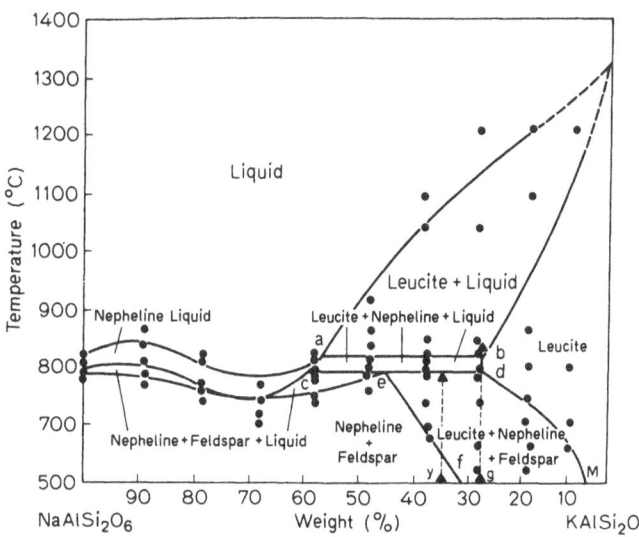

Fig. 13.3. Phase diagram of the system $KAlSi_2O_6$–$NaAlSi_2O_6$ under 1 kb P_{H_2O}. (After Fudali 1963)

leucite incorporates up to 28 wt.% $NaAlSi_2O_6$. The amount of solid solution increases as the water vapor pressure decreases. Below the solidus, sodic leucite breaks down to nepheline and K-feldspar as the temperature is lowered. Breakdown of sodic leucite below the solidus continues until either leucite is completely converted to nepheline and feldspar, or the remainder is so poor in sodium that it becomes stable. Fudali concluded that the subsolidus breakdown of leucite is responsible for the formation of pseudoleucite. He criticized the view of Bowen and Ellestad and considered that if pseudoleucite formation is related to solid–liquid reaction, the partially replaced crystals should be replaced from the rims inward; however, the reverse case was actually observed by him. Based on experimental observations he even thought that presence of viscous liquid coexisting with leucite may inhibit the formation of pseudoleucite (Fudali 1963; p. 1110).

The pseudoleucite-bearing rocks from Yukon were studied by Templeman-Kluit (1969), who did not favor either the subsolidus breakdown or the reaction theory of Bowen and Ellestad, but considered a combination of both theories for the formation of pseudoleucite.

Watkinson (1973) studied the pseudoleucites from Lakner Lake and Prairie Lake in Ontario, where they are associated with nepheline and feldspar. He supported the hypothesis of Fudali (1963), and concluded that these pseudoleucites were produced by the breakdown of sodic leucite at low P_{H_2O}.

Taylor and MacKenzie (1975) studied the undersaturated portion of the nepheline-kalsilite-SiO_2 system in connection with the genesis of pseudoleucite (also see Chap. 14.1). They heated three batches of the same mixture ($Ne_{25}Ks_{45}Qz_{30}$) at 840 °C and 2 kb P_{H_2O} and cooled them under three different conditions to produce leucites of three different textures (Fig. 13.4). When they cooled the pressure vessel containing the sample capsule with compressed air there was formation of leucite crystals, coexisting with glass and vapor. These leucites have exsolved phases of possibly analcitic composition. The bulk compositions of the leucites indicate that they are solid solutions, containing 11 wt.% of the soda-leucite molecule. When they cooled the pressure vessel more slowly, allowing both pressure and temperature to fall together, they were able to produce zoned leucite (Fig. 13.4) coexisting with glass and vapor. The rim of these crystals has compositions similar to that of the surrounding glass, which are equivalent to that of many pseudoleucites. Taylor and MacKenzie (1975) produced greatly exsolved leucites by cooling the pressure vessels first in air or in the

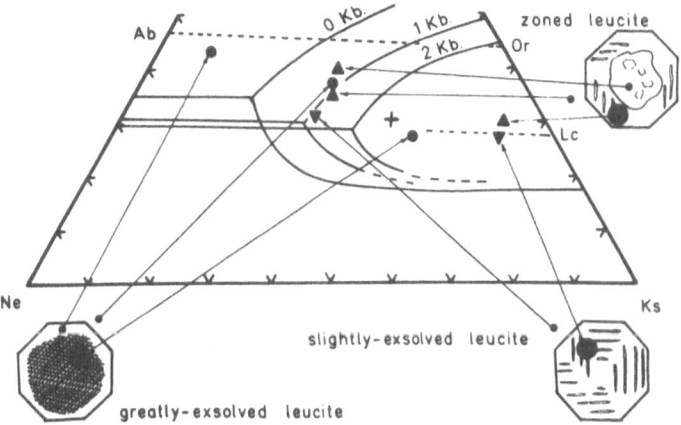

Fig. 13.4. Leucites of different textures formed by cooling under three different conditions. (After Taylor and MacKenzie 1975)

furnace from 840 °C to 630 °C and after holding it at 630 °C and 2 kb P_{H_2O} for 2 to 4 hrs, they cooled the pressure vessel in the final stage. Greatly exsolved leucites formed this way have a rim of compositions similar to that of analcite. The cores of these grains contain a mixture of leucite and analcite, whereas the rims contain 33 wt.% soda leucite.

Based on their experimental evidence Taylor and MacKenzie (1975) considered that the original leucite which crystallized from the magma was probably rich in sodium, but subsolidus reaction between leucite and sodium-rich hydrous liquid or the vapor surrounding the crystals might have resulted in exchange of potassium and sodium ions. Such a reaction should also cause a breakdown in the leucite structure. Taylor and MacKenzie criticized the hypothesis of Bowen and Ellestad on the ground that the solid–liquid reaction process would not explain the preservation of beautiful morphology of the crystals, which are pseudomorph after leucite.

Seki and Kennedy (1964) and later Scarfe et al. (1966) studied the subsolidus breakdown of leucite to kalsilite and K-feldspar (also see Chap. 14.1). The formation of pseudoleucite, containing intergrowth of K-feldspar and soda-poor kalsilite, may be possible due to subsolidus breakdown of leucite at elevated pressures. However, except for pseudoleucites of Synr, N. Baikal, Siberia, U.S.S.R. (Perchuk and Ryabchikov 1968) such an intergrowth of soda-poor kalsilite and K-feldspar in pseudoleucite is not known. The pseudoleucite-bearing rocks of Synr are ultrapotassic syenites, which are low in sodium, thus during the

formation of pseudoleucites, subsolidus ion exchange of K and Na between matrix and leucite crystals was not possible (Taylor and Mac-
Kenzie 1975).

Yagi and Gupta (1977) noticed the following characteristic features
of pseudoleucite:

1. Host rocks are generally sub-volcanic rocks, such as porphyritic
   tinguaite, probably rich in Na-bearing fluids in the later stage of
   crystallization.
2. Pseudoleucite has commonly well-preserved original crystal form
   of leucite.
3. Pseudoleucite is generally composed of fairly homogeneous aggregates of potash feldspar and nepheline.

Based on these observations and experimental evidence, Yagi and
Gupta (1977) proposed the following two-stage model for the genesis
of pseudoleucite.

First on ordinary leucite crystallizing out from tinguaitic magma is
transformed into "soda leucite" through the K $\rightleftharpoons$ Na substitution by
the Na-rich fluid in the later stage of crystallization. Later, this "soda
leucite" will break down to the aggregates of potash feldspar and
nepheline at subsolidus temperatures, retaining the original form of
leucite.

Some pseudoleucites with distinct zoned structure around fresh
leucite core in the lava flows may be formed by the solid-liquid reaction as advocated by Bowen and Ellestad (1937), but the genesis of
most of the well-shaped pseudoleucite found in the subvolcanic rocks
can be explained as the subsolidus break-down of secondary "soda
leucite".

Chapter 14 Study of Leucite-Bearing Systems Under Different Pressures and the P_{H_2O}-T Stabilities of Phlogopite, Potassium-Rich Richterite, and Kaersutite

14.1 Leucite-Bearing System Under Different Pressures

The univariant reaction, leucite ⇌ K-feldspar + kalsilite was studied by Scarfe et al. (1966) (Fig. 14.1), who also investigated the P_{H_2O}-T stability of leucite solid solutions containing 10% and 20% analcite. Their data show that leucite is unstable at high pressures, and at a given temperature the breakdown of leucite containing analcite takes place at a lower pressure than that of its pure synthetic equivalent. The study of Scarfe et al. explains the absence of leucite under plutonic conditions.

Schairer and Bowen (1935) and Schairer (1950) established that sanidine melts incongruently to leucite and a silica-rich liquid, under atmospheric pressure and there is a large field of leucite in the system nepheline-kalsilite-SiO_2 (Fig. 7.1). A minimum melting point occurs in the system at $Ne_{52}Ks_{15}Qz_{33}$ (see discussion in Chap. 7.1). Later study by Tuttle and Bowen (1958) (Fig. 14.2) and Hamilton and MacKenzie (1965) (Fig. 14.3) showed that under partial pressure of water, the field of leucite, produced as an incongruent melting phase of sanidine, decreases. Hamilton and MacKenzie noted that although the temperature of the minimum melting point is lowered from 1050° (established under atmospheric pressure) to 750 °C under 1 kb P_{H_2O}, its composition ($Ne_{50}Ks_{19}Qz_{31}$) is very similar to that found by

Fig. 14.1. Breakdown of leucite to K-feldspar and kalsilite under pressures. (After Scarfe et al. 1966). For a previous version of this breakdown curve see Seki and Kennedy (1964)

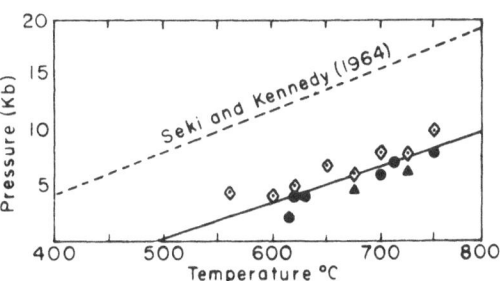

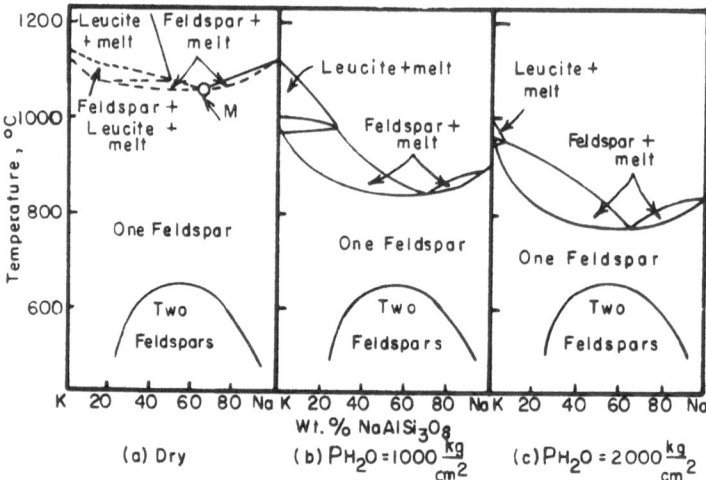

Fig. 14.2. Decrease in the field of leucite as an incongruent melting phase of sanidine. (After Tuttle and Bowen 1958)

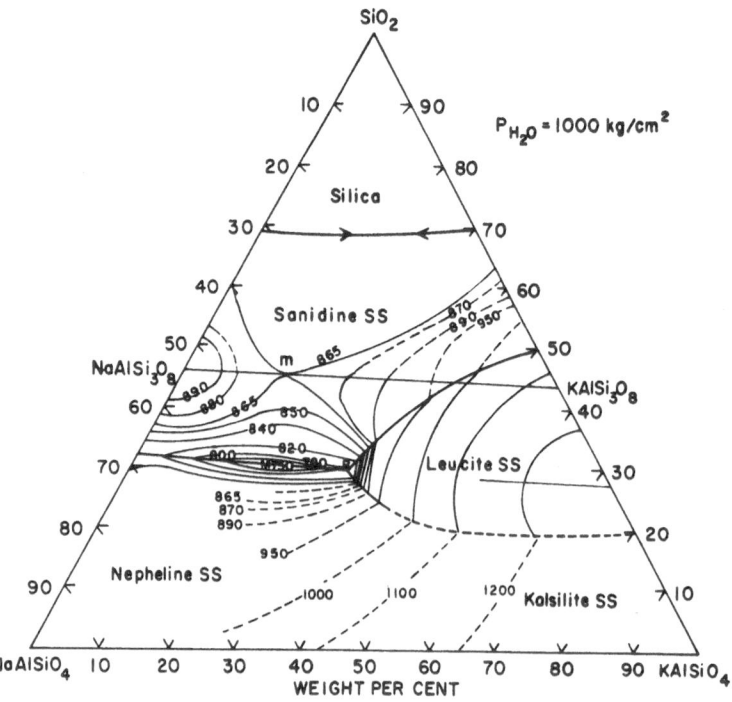

Fig. 14.3. Phase diagram of the system nepheline-kalsilite-SiO$_2$ at different temperatures under 1 kb P$_{H_2O}$. (After Hamilton and MacKenzie 1965)

Schairer (1950). Taylor and MacKenzie (1975) also studied the silica-undersaturated part of the system, nepheline-kalsilite-SiO_2 under 2 kb P_{H_2O} (Fig. 14.4) and showed that the minimum melting point occurs at $Ne_{51}Ks_{20}Qz_{29}$ and $710° \pm 7$ °C. Comparison of Figs. 7.1 and 14.4 shows a significant reduction of the leucite field under 2 kb P_{H_2O}. Morse (1968) studied the system nepheline-kalsilite-SiO_2 at 5 kb P_{H_2O} (Fig. 14.5) and showed complete absence of the field of leucite. Thus at a pressure between 2 and 5 kb P_{H_2O} leucite disappears completely. Morse also established a very narrow field of analcite$_{ss}$ between the fields of nepheline$_{ss}$ and sodic feldspar. He also noted the presence of a eutectic (E) and a reaction point R (Fig. 14.5).

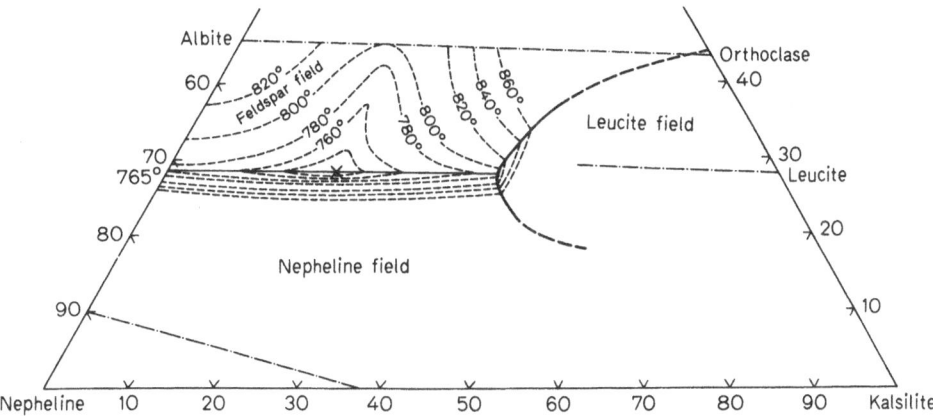

Fig. 14.4. Phase diagram of the system nepheline-kalsilite-SiO_2 at different temperatures under 2 kb P_{H_2O}. (After Taylor and MacKenzie 1975)

Incongruent melting of sanidine (Fig. 14.6) was studied up to 40 kb under dry condition by Lindsley (1966), who found that the field of leucite + liquid persists up to 19 ± 1 kb at 1440 °C; above this pressure sanidine melts congruently. It can be seen from the negative slope that the melting of leucite produces a liquid of higher density. Lindsley also studied a portion of the $KAlSiO_4$–SiO_2 system under dry pressure. His data (Fig. 14.7) show the existence of two singular points S_1 (obtained from Fig. 14.6) and S_2. I is the invariant point, produced by the intersection of the following curves: kalsilite + leucite = liquid, leucite = kalsilite + sanidine (Scarfe et al. 1964) and kalsilite + sanidine = liquid.

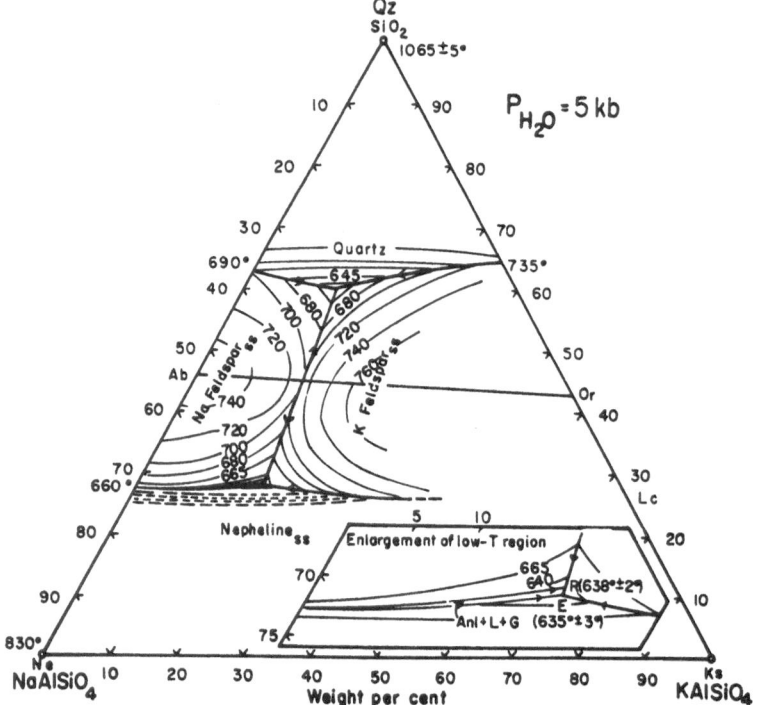

Fig. 14.5. Phase diagram of the system nepheline-kalsilite-SiO_2 at different temperatures under 5 kb P_{H_2O}. (After Morse 1969)

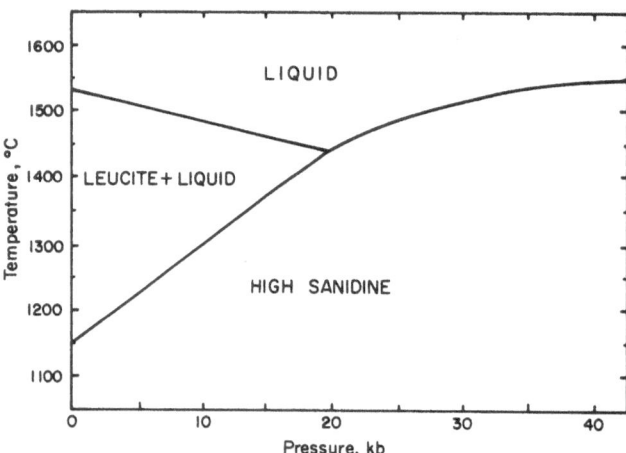

Fig. 14.6. High pressure study of the incongruent melting of sanidine under dry conditions. (After Lindsley 1966)

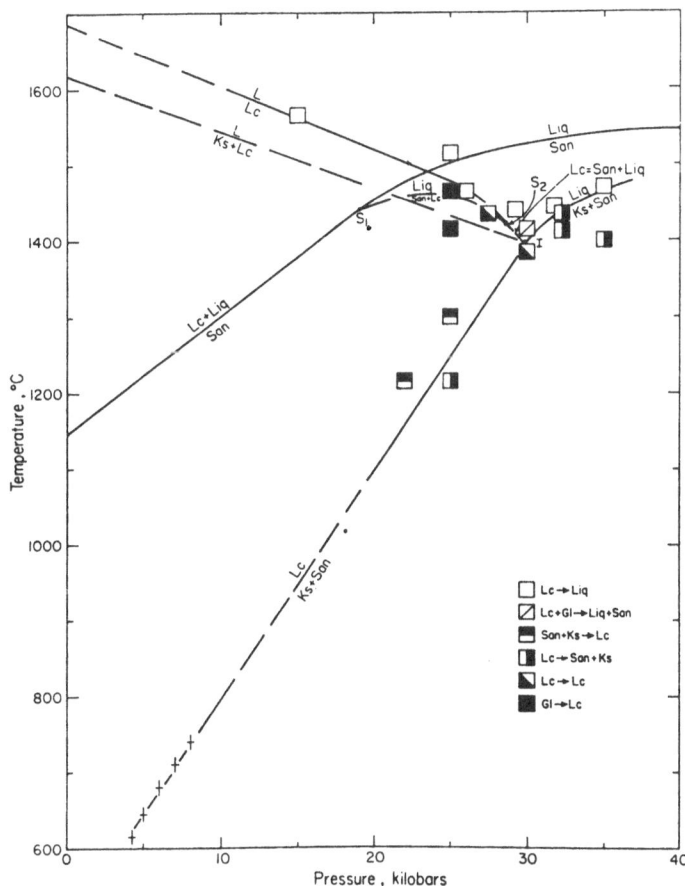

Fig. 14.7. P–T diagram of a portion of the system KAlSiO$_4$–SiO$_2$ under dry pressures. (After Lindsley 1966)

14.2 The System KAlSiO$_4$–Mg$_2$SiO$_4$–SiO$_2$–H$_2$O

The bulk compositions of leucite-bearing mafic and ultramafic volcanic rocks from West Kimberley, New South Wales, the Leucite-Hills, and southern Spain can be closely approximated by the system CaMgSi$_2$O$_6$–Mg$_2$SiO$_4$–KAlSiO$_4$–SiO$_2$–H$_2$O. The whole system has not been studied systematically, but the join KAlSiO$_4$–Mg$_2$SiO$_4$–SiO$_2$–H$_2$O has been investigated up to 3 kb in presence of excess water by Luth (1967). He plotted the P$_{H_2O}$–T stability of 35 univariant reactions (Table 14.1) as shown in Fig. 14.8, which indicates the presence of nine quaternary

Table 14.1. Invariant points (Roman numerals) and univariant curves, or reactions (Arabic numerals) in the system $KAlSiO_4-Mg_2SiO_4-SiO_2-H_2O$. (After Luth 1967)

I. Ph + Fo + Ok + Lc + L + V

    1 Fo + Lc + V = L[b]
    2 Fo + Lc + Ok + V = L
    3 Ph = Fo + Ok + Lc + V
    4 Ph = Fo + Ok + L + V[a]
    5 Ph + Ok + V = Fo + L[a]
    6 Ph = Fo + L + V[a]
    7 Ph + Ok + Lc + V = L
    8 Ph + Lc + V = Fo + L[b]
    9 Ph + Lc + Fo + V = L[b]
    10 Ph + L = Fo + Lc + V[b]
    11 Ph + Lc + V = L[b]
    29 Ph + Lc = Fo + Ok + L

II. Ph + Fo + Or + Lc + L + V

    12 Or + Fo + V = Lc + L
    13 Ph + Or = Fo + Lc + V
    14 Or + Ph + V = Lc + L[c]
    15 Or + Ph + Lc + V = L[c]
    16 Or + Ph + V = L[c]
    17 Ph + L = Or + Fo + V
    10 Ph + L = Fo + Lc + V
    31 Ph + Or = Lc + Fo + L

III. Ph + Fo + En + Or + L + V

    18 Or + En + V = Fo + L
    19 Ph + En = Or + Fo + V
    20 Ph + En = Fo + L + V
    21 Ph + L = Or + En + V
    17 Ph + L = Or + Fo + V
    32 En + Or + Ph = Fo + L

IV. Fo + Lc + Or + En + L + V

    22 Lc + En + V = Fo + L
    23 Or + En + V = Lc + L
    12 Or + Fo + V = Lc + L
    18 Or + En + V = Fo + L
    30 Or + Fo = Lc + En

V. Ph + Or + Q + En + L + V

    24 Or + En + Q + V = L
    25 Ph + Q = Or + En + V
    26 Ph + Or + Q + V = L
    27 Ph + Q + V = En + L
    21 Ph + L = Or + En + V
    28 Ph + Q = Or + En + L

VI. En + Pen + Lc + Fo + L + V

    33 En = Pen
    22a Lc + En + V = Fo + L
    22b Lc + Pen + V = Fo + L

VII. En + Pen + Lc + Or + L + V

    33 En = Pen
    23a Or + En + V = Lc + L
    23b Or + Pen + V = Lc + L

VIII. Q + Tr + Or + En + L + V

    34 Q = Tr
    24a Or + En + Q + V = L
    24b Or + En + Tr + V = L

IX. Ok + Ks + Ph + Lc + L + V

    35 Ks = Ok
    7a Ph + Ok + Lc + V = L
    7b Ph + Ks + Lc + V = L

[a] Related by singular point on the univariant curve involving Ph, Fo, Ok, L, and V
[b] Related by singular points on the univariant courves involving Ph, Fo, Lc, L, and V
[c] Related by the singular points on the univariant curves involving Ph, Or, Lc, and V

All reactions are written in such a way that the high temperature assemblages is on the right side of the reaction symbol, except for reaction 10. See also List of Abbreviations

invariant points (hexagonal symbol), where six phases are in equilibrium. Figure 14.8 also shows the presence of four singular points, generated by the meeting of the curves 14, 15, and 16; 4, 5, and 6; 8, 9, and 11; and 1, 9, and 10.

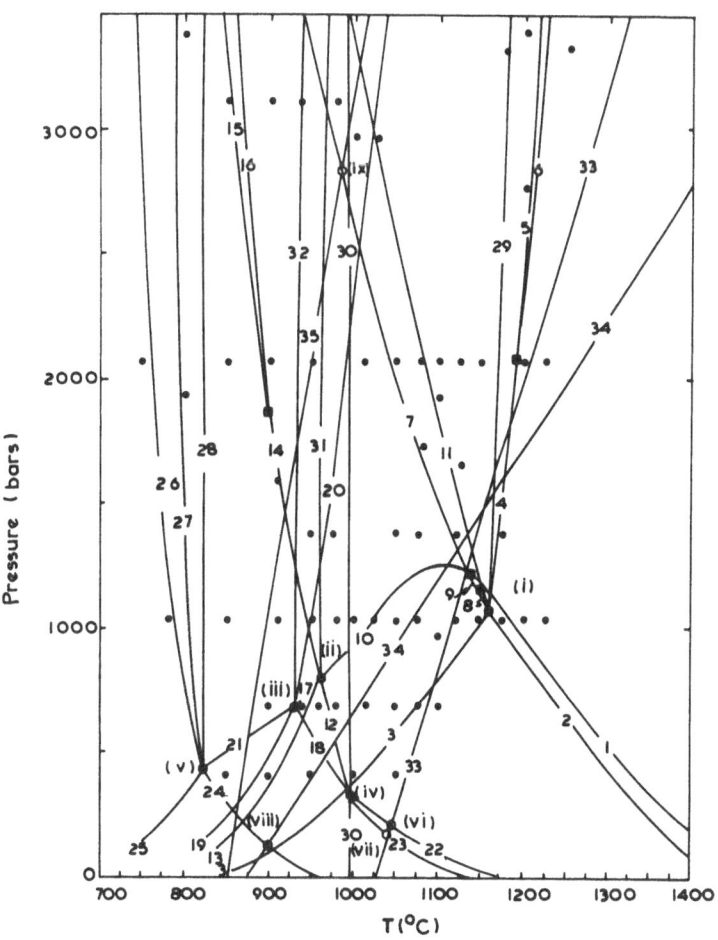

Fig. 14.8. P–T projection of inferred invariant and univariant equilibria in the system KAlSiO$_4$-Mg$_2$SiO$_4$-SiO$_2$-H$_2$O. The *lower case Roman numerals* and *Arabic numerals* refer to invariant and univariant equilibria, given in Table 14.1. The position of v is from Wones (quoted in Luth 1967). *Filled hexagonal symbols* refer to quaternary invariant points. *Open hexagonal symbols* indicate quaternary invariant points, generated by polymorphism of KAlSiO$_4$, MgSiO$_3$, or SiO$_2$. *Filled squares* refer to postulated singular points. *Solid circles* indicate data points. (After Luth 1967)

On the basis of Fig. 14.8, Luth (1967) constructed a series of isobaric polythermal, polybaric polythermal, and polybaric isothermal
diagrams (Fig. 14.9) with inferred phase relations. These diagrams are
produced by projection of saturation surfaces on the anhydrous base
(Mg_2SiO_4–$KAlSiO_4$–SiO_2 plane) of the tetrahedron from the H_2O apex.
The saturation surfaces are the areas, where a vapor phase is in equilibrium with a water-saturated silicate melt. The P–T conditions of various
inferred phase diagrams (Fig. 14.9a–i) can be obtained from Fig. 14.9j.

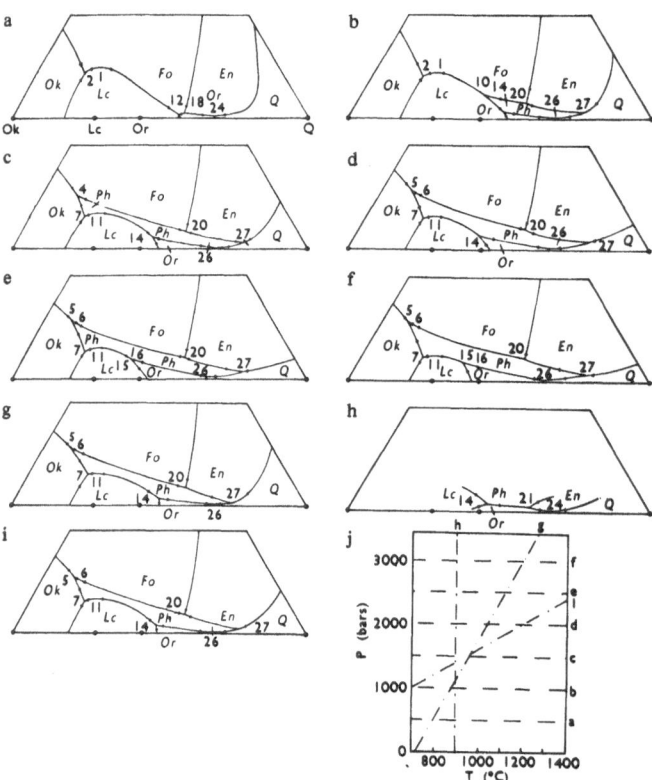

Fig. 14.9a–j. Inferred phase relations on the saturation surface. Projection of the
saturation surface on the anhydrous base of the tetrahedron from the H_2O apex.
a Isobaric-polythermal, 500 bars (a of Fig. 14.9j). b Isobaric-polythermal, 1000
bars (b of Fig. 14.9j). c Isobaric-polythermal, 1500 bars (c of Fig. 14.9j). d Isobaric-polythermal, 2000 bars (d of Fig. 14.9j). e Isobaric-polythermal, 2500 bars
(e of Fig. 14.9j). f Isobaric-polythermal, 3000 bars (f of Fig. 14.9j). g Polybaric-
polythermal (g of Fig. 14.9j). h Polybaric-isothermal, 900 °C (h of Fig. 14.9j).
i Polybaric-polythermal (i of Fig. 14.9j). j Pressure-temperature orientation for
Fig. 14.9a–i. (After Luth 1967)

It should be pointed out here that there are some discrepancies in the results between Figs. 14.8 and 14.9 as presented by Luth (1967). For example, according to Fig. 14.8 (reaction 21, Table 14.1) there should be a field of phlogopite at 500 bar in the kalsilite-rich side of Fig. 14.9a. Figure 14.8 also does not suggest the existence of reaction 14 at 2000 bar as shown in Fig. 14.9d.

Figure 14.9a–i show that at pressures at or below 2 kb there are liquids of two different compositions. One is silica-undersaturated and coexists with orthorhombic kalsilite, leucite, forsterite (or phlogopite), and vapor and the other is silica-saturated and is in equilibrium with K-feldspar, quartz, enstatite (or phlogopite), and vapor. According to Luth (1967), these liquids are simplified equivalents of phonolite, rhyolite, and trachyte respectively. Comparison of Fig. 14.9a with Fig. 14.9b–i shows that at or above 1 kb phlogopite appears as a phase in this system, whereas near 500 bar it completely disappears by reaction with a liquid. Bowen (1928) considered that a reaction between mica and a mafic liquid have played an important role in the genesis of potassium-rich undersaturated liquid; this will be discussed in detail in a subsequent chapter. Figure 14.9a–i show that forsterite is a liquidus phase in a wide range of compositions. With further variation of bulk compositions, leucite, K-feldspar, enstatite and/or phlogopite may join olivine in the course of equilibrium or fractional crystallization. At a pressure of less than 300 bar (near the surface) the resulting crystal assemblage should consist of forsterite, leucite, and enstatite with siliceous interstitial liquid. At slightly higher water pressures, forsterite would cease to exist (depending on bulk compositions), and phlogopite and leucite, phlogopite and enstatite, or phlogopite and K-feldspar should be stable. In the case of natural leucite-bearing rocks, although K-feldspar is an accompanying phase, enstatite is absent.

On the basis of Fig. 14.9 Luth (1967) considered that a magma at depth initially precipitating olivine should cool with resorption of olivine and formation of mica and then pyroxene. These early differentiates should be present as xenoliths and react with liquid. He referred to the existence of phlogopite-bearing ultramafic xenoliths in the potassium-rich lavas of the Bufumbira region (Holmes and Harwood 1937), to support his conclusion. Luth also considered that the magma of this region cooled for a long period at moderate pressure to produce differentiates of the biotite pyroxenite-peridotite series. which was followed by their emplacement at shallow depths. Because of the presence of leucite (pseudoleucite), olivine and clinopyroxene and absence of phlogopite, he considered that the rocks of Shonkin Sag Laccolith were produced under low water pressures at shallow depths. He suggested that the presence of mica along with leucite, clinopyroxene, and amphibole is indicative of the fact that the rocks of West Kimberley were produced by a long cooling process of a potassic magma at a moderate depth. From the rarity of forsterite and presence of large amounts of phlogopite, he considered that the rocks of the Leucite Hills were produced at a relatively greater depth.

14.3 Phlogopite Stability

Close genetic relations between leucite-bearing mafic and ultramafic rocks and mica-bearing pyroxenites and peridotites (included within these lavas as xenoliths) were pointed out by Holmes and Harwood (1937). The mica associated with these potassic rocks is usually phlogopite. Yagi and Matsumoto (1966), Cundari and Le Maitre (1970), and Edgar et al. (1976) also considered that potassium-rich mafic and ultramafic rocks are represented at depth by phlogopite-bearing ultramafic rocks. Stability of phlogopite at depth as a source of potassium is therefore of great significance.

Melting relations of phlogopite under vapor-present and vapor-absent conditions were studied by Yoder and Kushiro (1969) up to 37.5 kb. Their results are summarized in Fig. 14.10, which shows the P_{H_2O}–T stability of the following reactions:

A: phlogopite + vapor = forsterite + liquid.
B: beginning of melting of the vapor-absent assemblage, forsterite + phlogopite + kalsilite (orthorhombic kalsilite) + liquid.
C: phlogopite = forsterite + liquid (vapor-absent condition).
D: minimum liquidus in the presence of a vapor phase in the phlogopite + H_2O join. Position of D varies with the H_2O content of the system.
L: forsterite + orthorhombic kalsilite + leucite + vapor = phlogopite (Luth 1967; Wones 1967).
G: phlogopite = forsterite + leucite + kalsilite + liquid.

The curves above the invariant point involving phlogopite, forsterite, orthorhombic kalsilite, leucite, liquid, and vapor at about 1160 °C and 1 kb indicate that the breakdown of phlogopite occurs according to the reaction phlogopite + vapor = forsterite + orthorhombic kalsilite + liquid. Under vapor-absent equilibrium conditions phlogopite melts incongruently according to reaction, phlogopite = forsterite + leucite + orthorhombic kalsilite + liquid (up to 1.7 kb), after which the reaction becomes phlogopite = forsterite + liquid (Fig. 14.10).

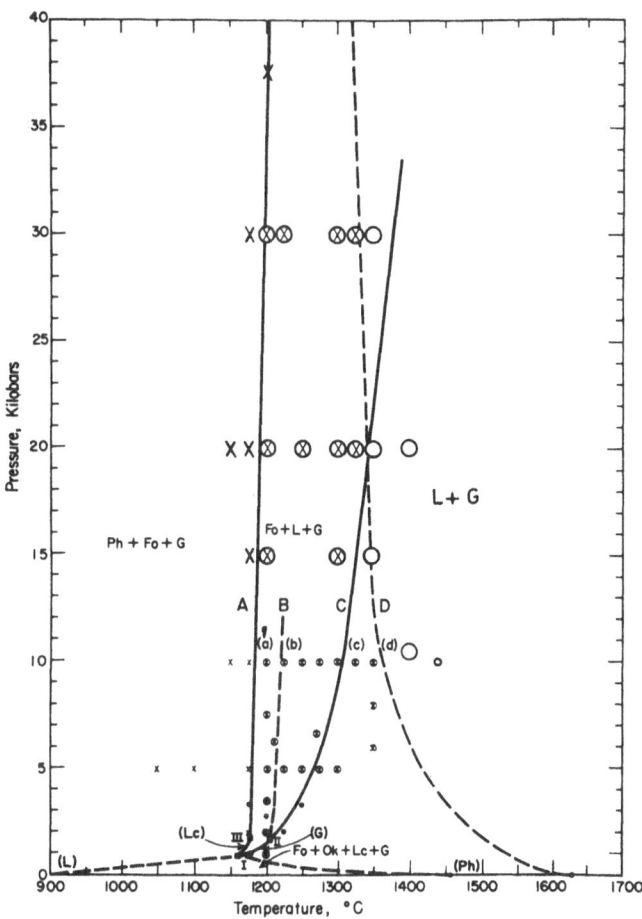

Fig. 14.10. P–T diagram for compositions in the join $K_2O \cdot 6\, MgO \cdot Al_2O_3 \cdot 6\, SiO_2$-$H_2O$. The *solid squares* mark an invariant point I and two singular points II and III. X = crystal + gas; circle with an X = crystal + liquid + gas; circle = liquid + gas. The symbols are only relevant to *curves A* and *D*. *Curve A* is the maximum stability of phlogopite in the presence of a gas phase. *Curve B* marks the beginning of melting of the gas absent assemblage Ph + Fo + Ks (or Ok) + L, and *curve C* is the maximum stability of phlogopite in the absence of a gas phase. *Black dots* are the data points of Luth (1967). (After Yoder and Kushiro 1969). For other details see text and List of Abbreviations

The importance of the presence or absence of a vapor phase with respect to the stability of leucite and kalsilite can be recognized from the isobaric (10 kb) diagram of the system $K_2O \cdot 6 \, MgO \cdot Al_2O_3 \cdot 6 \, SiO_2$-$H_2O$, which was studied by Yoder and Kushiro (1969) at various temperatures (Fig. 14.11). These data show that in presence of small amounts of water (4 wt.% or lower) leucite and kalsilite may be stable up to 10 kb even at 1200 °C (Fig. 14.10).

The P_{H_2O}-T stability of titan-phlogopite [$K_2Mg_4TiAl_2Si_6O_{20}(OH)_4$ was studied by Forbes and Flower (1974) (Fig. 14.12)]. Comparison of Figs. 14.10 and 14.12 suggests that while the breakdown temperatures of phlogopite in presence of gas at 25 and 30 kb are 1170° and

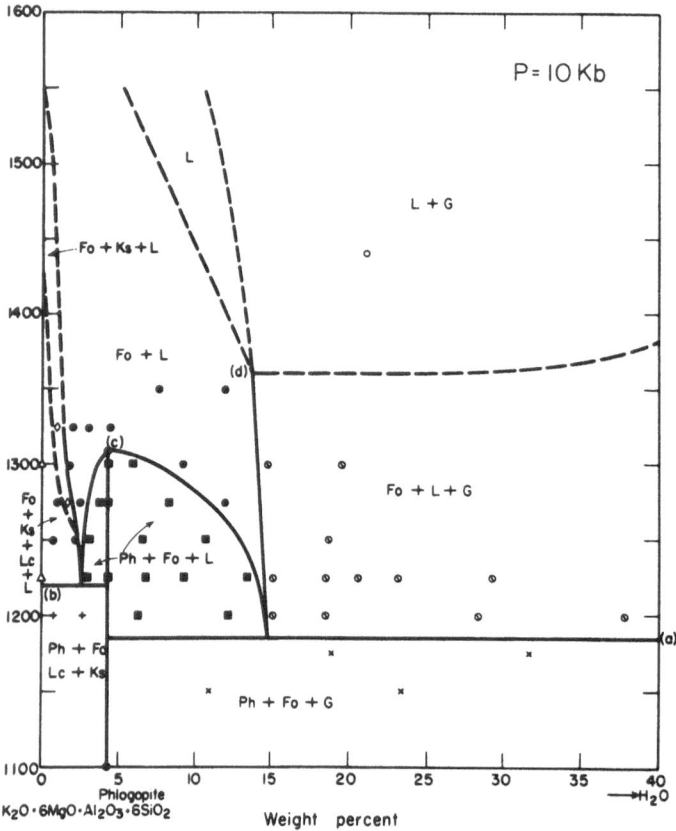

Fig. 14.11. Phase diagram of the system $K_2O \cdot 6 \, MgO \cdot Al_2O_3 \cdot 6 \, SiO_2$-$H_2O$ under isobaric polythermal conditions. (After Yoder and Kushiro 1969). See also List of Abbreviations

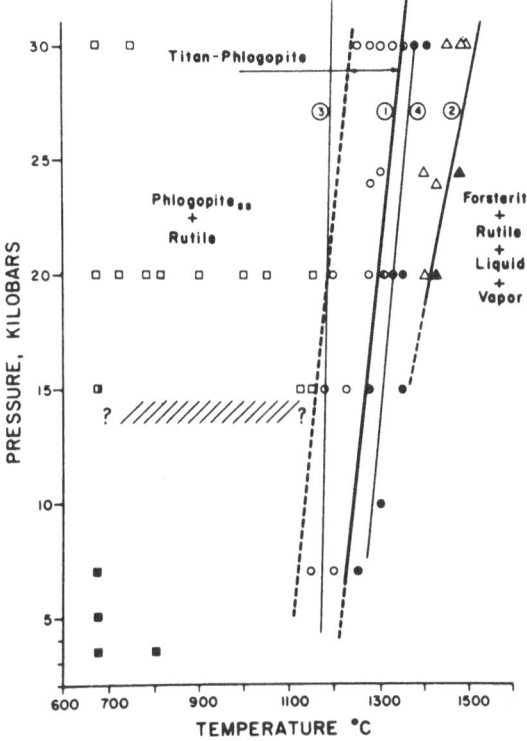

Fig. 14.12. Melting curves for titan-phlogopite + excess vapor *(curve 1)* and titan-phlogopite in the absence of vapor *(curve 2)*. Decomposition curves for phlogopite in the presence and absence of vapor are shown by *curves 3* and *4* respectively. The *dashed line* delimits the synthesis range of titan-phlogopite as opposed to phlogopite solid solution + rutile or phlogopite + sanidine + rutile. Symbols – vapor-present runs: *filled squares* = Ph + Sa + Ru; open squares = Ph_{ss} + Ru; *open circles* = Ti-Ph + vapor; *filled circles* = Fo + Ru + L + vapor. Vapor-absent runs: *open triangles* = Ti-Ph; *filled triangles* = Fo + Ru + L ± vapor. (After Forbes and Flower 1974). Ru and Ti-Ph designate rutile and titan-phlogopite respectively. See also List of Abbreviations

1175 °C respectively, the upper stability limit of titan-phlogopite at the same pressures is 1200° and 1220 °C respectively. The above observation suggests that titan-phlogopite is stable up to higher temperatures under similar pressure conditions than phlogopite without TiO_2. However, in their study of some phlogopites, crystallizing from a starting material, similar to that of a synthetic mafurite, Edgar et al. (1976) noted that at a given temperature with increasing pressure the TiO_2 content of phlogopite, coexisting with ilmenite, pyroxene and liquid systematically decreased, suggesting that the solubility of TiO_2 in phlogopite in more complex systems is more limited at high pressures.

Modreski and Boettcher (1972) studied the stability of phlogopite in presence of enstatite between 2 and 35 kb. The stability of phlogopite under vapor-absent and vapor-present conditions is given in Figs. 14.13 and 14.14. As orthopyroxene and olivine are important phases in the upper mantle, the results of Modreski and Boettcher may be a good approximation of the stability of phlogopite in the upper mantle. A comparison of the stability of phlogopite with that of magnesium iron trioctahedral mica (Eugster and Wones 1962; Luth 1967; Wones and Eugster 1965; Markov et al. 1966, 1968) helped Modreski and Boettcher to estimate that phlogopite containing 7 wt.% FeO (the amount of iron expected in mantle phlogopite) should have a stability $50° \pm 25$ °C lower than that of iron-free phlogopite. From the experimental results of Rutherford (1969), they concluded that phlogopite

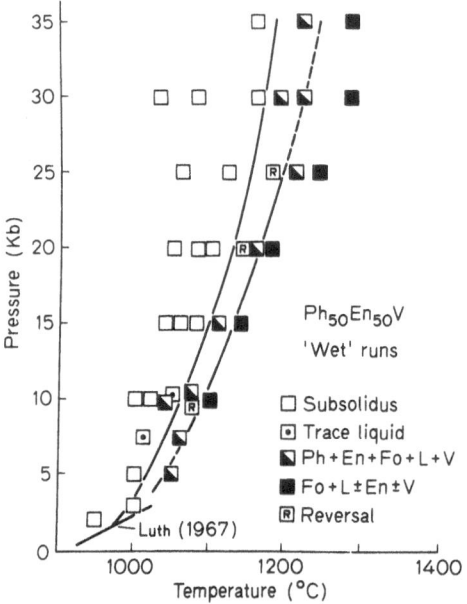

Fig. 14.13. Stability of phlogopite in the system phlogopite-enstatite under vapor-present conditions, defining the reaction, phlogopite + enstatite + vapor ⇌ forsterite + liquid. The *left-hand curve* indicates the beginning of melting; the *right-hand curve* (*dashed*, where extrapolated) indicates the disappearance of phlogopite. The curve of Luth (1967) for the melting of phlogopite + enstatite + vapor is also shown. (After Modreski and Boettcher 1972)

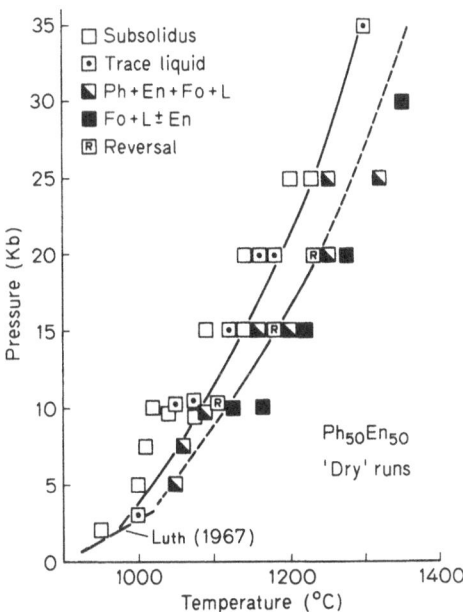

Fig. 14.14. Stability of phlogopite in the system phlogopite-enstatite under vapor-absent conditions, defining the reaction, phlogopite + enstatite ⇌ forsterite + liquid. The *left-hand curve* indicates the beginning of melting. The *right-hand curve* (*dashed*, where extrapolated) indicates the disappearance of phlogopite. The curve of Luth (1967) for melting of phlogopite + enstatite + vapor is also shown. (After Modreski and Boettcher 1972)

containing sodium phlogopite (Carman 1969) has much lower P_{H_2O}-T stability.

Modreski and Boettcher (1973) also studied the stability of phlogopite in presence of diopside at different temperatures under both dry and wet conditions (Fig. 14.15). When the P_{H_2O} stabilities of phlogopite + diopside (Fig. 14.15) and phlogopite + enstatite (Fig. 14.14) are compared it is noted that at a given pressure the former assemblage is stable to higher temperature conditions than the latter. Modreski and Boettcher (1972) also compared the stability of phlogopite + enstatite with the P-T conditions of shield and oceanic geotherms and the solidi of hydrous and anhydrous peridotites (Fig. 14.16). From their study they concluded that phlogopite should be unstable below about 75 km in the regions of the mantle with a high geothermal gradient. Melting of phlogopite at this depth may result in the formation of basaltic magma, whereas in the regions of low geothermal gradient, phlogopite may persist to depths greater than 175 km.

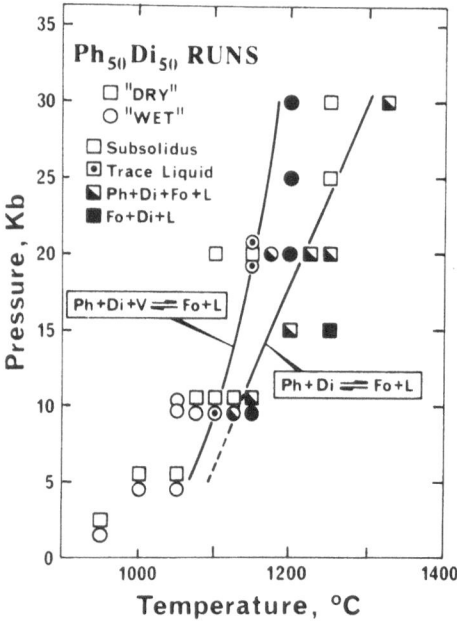

Fig. 14.15. P–T projection of vapor-absent (dry) and vapor-present (wet) runs on the bulk composition phlogopite$_{50}$diopside$_{50}$ (wt.%). Reactions, phlogopite + diopside $\rightleftharpoons$ forsterite + liquid and phlogopite + diopside + vapor $\rightleftharpoons$ forsterite + liquid, are shown. (After Modreski and Boettcher 1973)

14.4 Potassic Richterite and Kaersutite as Possible Sources of Potassium in the Upper Mantle

14.4.1 Stability of Potassic Richterite

Potassic richterite is often noted in the leucite-bearing lavas of West Kimberley, Leucite Hills, New South Wales, and East Eifel. Table 3.9 indicates that in these richterites the K_2O content can be as high as 2.5%. Because of high K_2O and K_2O/Na_2O ratio (>1), this amphibole may be considered as an important source of potassium in the upper mantle. Amphibole as a possible supplier of K_2O in the mantle was suggested by Oxburgh (1964). Occurrence of potassic richterite in diopside-phlogopite nodules from the Wesselton kimberlite, Kimberley (Erlank and Finger 1970) and in mica nodules in a south African kimberlite (Aoki 1974) further supports this view. In case of the last-mentioned area, K_2O content of the richterite is as high as 4.98%

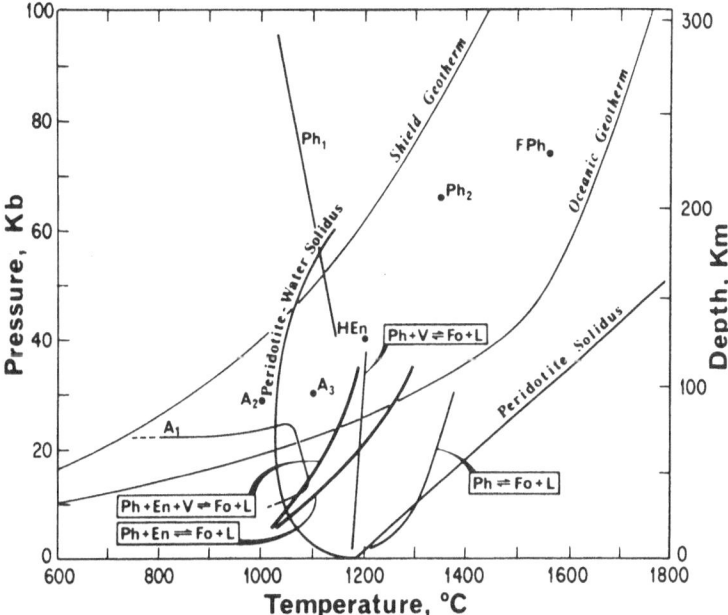

Fig. 14.16. P–T projection of some reactions related to the stability of hydrous minerals in the mantle. Ph + En ⇌ Fo + L and Ph + En + V ⇌ Fo + L are from Modreski and Boettcher (1972). Ph ⇌ Fo + L ± V are from Yoder and Kushiro (1969). A_1 stability limit of amphibole in water-saturated alkali basalt; A_2' stability limit of amphibole in pyrolite. A_3 conditions at which potassic richterite is stable. *H En* minimum pressure at which hydroxylated orthoenstatite is stable; Ph_1 and Ph_2 stability limits of natural phlogopites. *FPh* conditions at which synthetic fluor-phlogopite is stable. Except for the curves Ph + En ⇌ Fo + L and Ph + En + V ⇌ Fo + L, all other data points and the stability curves are from other sources. (After Modreski and Boettcher 1972). See also List of Abbreviations

(Aoki 1974). The stability of richterite at different temperatures and pressures is therefore discussed below.

Kushiro and Erlank (1970) studied a potassic richterite in the presence of pure diopside up to 30 kb in presence of water. They found that under water-saturated conditions at 1100°, 1000 °C, and 30 kb, potassic richterite was stable and coexisted with diopside. The K_2O content of richterites in these runs varied from 4.6% to 5.7%. However, in the presence of a pyrope-rich garnet (containing 33 mol% grossularite), even at 20 kb and 1000 °C, it reacted to form omphacite, phlogopite, and minor orthopyroxene. The data of Kushiro and Erlank (1968) therefore suggest that in presence of garnet, potassic richterite should not be stable even at a depth of 60 km, whereas in presence of alumina-poor clinopyroxenes, it may survive even below 90 km. In their study of a synthetic melilite-nepheline leucitite under various P_{H_2O}-T conditions, Gupta et al. (1976) noted that potassic richterite has a lower P-T stability field than phlogopite and clinopyroxene (Fig. 15.3).

14.4.2 Stability of Kaersutite

Because of their high K_2O and TiO_2 contents, Yagi et al. (1975) considered that kaersutite may be a possible source of potassium and titanium in the upper mantle. They also described the occurrence of this mineral in a potassium-rich trachybasalt. Merrill and Wyllie (1975) noted that the K_2O and TiO_2 contents of kaersutite from a mineral breccia, collected from Kakanui, New Zealand, is as high as 2.2% and 4.46% respectively. It should be mentioned, however, that the K_2O/Na_2O ratio of kaersutite is < 1. Kaersutite from Kakanui is compositionally equivalent to an olivine nephelinite. Leucite-bearing rocks are not only high in K_2O and TiO_2, but they are often associated with olivine, nepheline, and clinopyroxene. Study of the P_{H_2O}-T stability of kaersutite at high

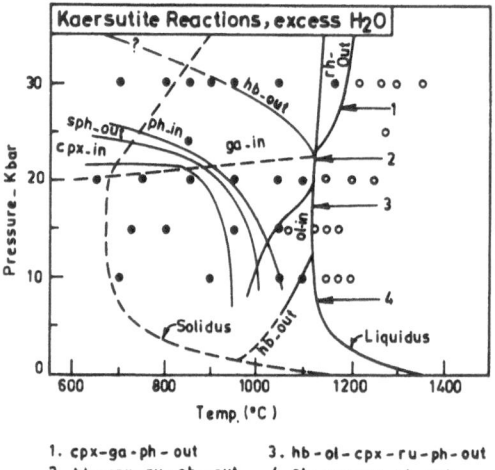

1. cpx-ga-ph-out   3. hb-ol-cpx-ru-ph-out
2. hb-cpx-ru-ph-out   4. ol-cpx-ru-ph-out

Fig. 14.17. Experiments with kaersutite megacryst (olivine nephelinite) plus excess water. *Light solid lines* indicate field boundaries. *Heavy solid line* denotes water-saturated liquidus contour. *Heavy dashed line* is estimated from the data of other workers (see Merrill and Wyllie 1975). Experiments are plotted with circles: *solid circles* denote runs whose products include crystals; *open circles* indicate crystal-free run products. (After Merrill and Wyllie 1975). See also List of Abbreviations

pressures as a possible source of K_2O, Na_2O, and TiO_2 in the mantle is therefore important.

Merrill and Wyllie (1975) studied the powdered sample of a kaersutite megacryst from Kakanui, New Zealand, under pressures between 10 and 30 kb and at temperatures between 650° and 1350 °C in presence of excess water. Their results are summarized in Fig. 14.17, which shows that below 20 kb, kaersutite melts incongruently near the solidus to produce amphibole of different composition, coexisting with minor amounts of sphene, rutile, and liquid. At temperatures just above the "cpx in" curve sphene reacts with liquid to produce clinopyroxene and liquid. Minor phlogopite ($<$ 1 vol %) appears just above the "sph out" curve and is present up to the liquidus. Above the "Ga in" boundary, kaersutite breaks down near the solidus to an assemblage of amphibole, clinopyroxene, pyrope-rich garnet ± rutile ± sphene ± phlogopite ± liquid. From the presence of sphene and rutile in the low temperature side, and rutile in the high temperature side of the diagram, it appears that titanium-rich amphiboles are stable only under low pressure environment. It is also apparent that the K_2O content of amphibole decreases in the region marked by the appearance of phlogopite. The run products (Table 2 of Merril and Wyllie 1975) also show that phlogopite has higher P–T stability than amphibole.

Merrill and Wyllie studied the same sample of kaersutite megacryst under pressure without adding water (Fig. 14.18). This mixture contains 0.9 wt. % water structurally bonded to amphibole. In this study, they noted the presence of amphiboles including acicular crystals of

Fig. 14.18. Experiments with kaersutite megacryst (olivine nephelinite) under H_2O-deficient (vapor-absent) conditions: 0.9 wt. % H_2O initially bonded to kaersutite crystal. Liquidus contour is extrapolated to meet water-saturated contour (from Fig. 1 near point 0.8 kb, estimated from the solubility data of other workers; see Merrill and Wyllie 1975). Solidus from Fig. 14.17. Phlogopite was observed from runs at 30 kb below 1200 °C. See also List of Abbreviations

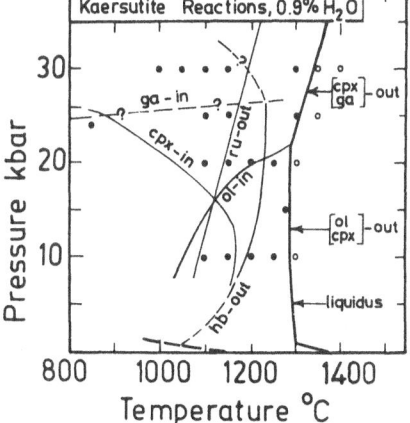

rutile in the runs at 30 kb, 1100°, and 1150 °C. Amphibole without rutile inclusions is stable at 1000° and 1050 °C at the same pressure. Amphibole field extends to even 1200 °C at 20 kb. These P–T conditions relating to amphibole stability are well above its field boundary, observed with excess water. Minor phlogopite occurs in runs at 30 kb, but not at the liquidus, and garnet is present only above 30 kb.

In a P–T diagram (Fig. 14.19) Merrill and Wyllie (1975) plotted various curves showing the maximum stability of amphiboles in water-saturated mafic liquids of variable compositions. Figure 14.19 shows that amphiboles crystallizing from silica-poor sodium-rich liquids have higher P–T stability than those crystallizing from silica-saturated liquid, and above 33 kb (corresponding to about 100 km) amphiboles are unstable.

Discussion in Chaps. 14.3 and 14.4 suggests that the P_{H_2O}–T stabilities of potassic richterite and kaersutite are much lower than that of phlogopite. While amphiboles do not survive below 100 km, phlogo-

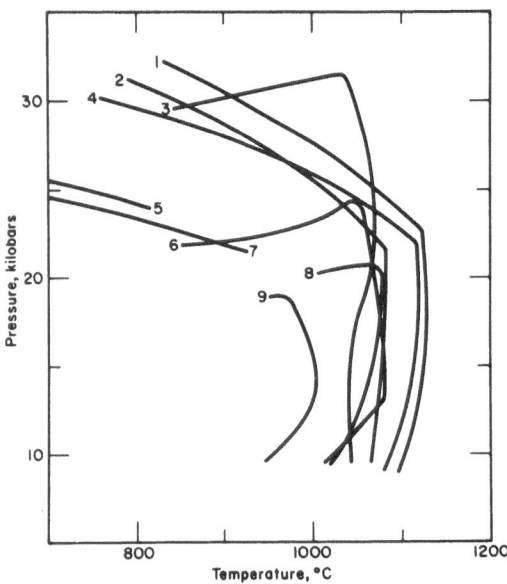

Fig. 14.19. Experimentally determined stability limits ("hb-out" field boundaries in equilibrium with water-saturated basaltic liquids. *1* Kaersutite megacryst; *2* kaersutite eclogite nodule; *3* brown hornblende mylonite; *4* olivine nephelinite; *5* high-alumina olivine tholeiite; *6* alkali olivine basalt; *7* "alkali olivine basalt", synthetic mixture; *8* olivine tholeiite; *9* quartz tholeiite. Curves *3* to *9* were determined by various workers. (After Merrill and Wyllie 1975)

pite may be stable up to a depth of 175 km (Modreski and Boettcher 1972). Because of its higher K_2O/Na_2O ratio, of the two amphiboles potassic richterite (rather than kaersutite) seems to be a preferable source of potassium (in addition to phlogopite) in the upper mantle.

Chapter 15 Study of Leucite-Bearing Rocks in Air and Under Various P–T Conditions

There have been several experimental investigations on leucite-bearing rocks to determine the nature of their parent magma. Some atmospheric studies of these rocks by Yagi and Matsumoto (1966), Sobolev et al. (1975) and Fyfe (quoted in Carmichael 1967) are discussed first. Results of investigation on synthetic and natural rocks under various P–T conditions are described later.

15.1 One-Atmospheric Studies on Leucite-Bearing Rocks

Yagi and Matsumoto (1966) made some heating experiments at 1 atm on a natural orendite (1), wyomingite (2), and a xenolith (3) included within the wyomingite, all from the Leucite Hills, Wyoming. The orendite contained modal leucite, sanidine, clinopyroxene, apatite, and calcite. The wyomingite consisted of leucite, diopside, and phlogopite with small amounts of apatite, calcite, and magnetite. The xenolith contain clinopyroxene and leucite. The liquidus temperatures of the rocks (1), (2), and (3) are as follows: 1302°, 1322°, and 1318 °C. On the basis of their thin section studies of these natural leucite-bearing rocks Yagi and Matsumoto considered that leucite invariably crystallized earlier, whereas in their experimental study in air, they noted that clinopyroxene was always the liquidus phase. This discrepancy was interpreted by them to be the result of the absence of water pressure in their investigation. Rocks (2)–(3) have liquidus temperatures, which are much higher than those of tholeiite and an alkali basalt (Tilley et al. 1965).

Sobolev et al. (1975) made some heating experiments in air on a wyomingite from the Leucite Hills, the mineralogical composition of which is the same as that mentioned above, and found the liquidus temperature to be 1320 °C. They made thermal experiments on fluid inclusions in phlogopite, diopside, and leucite to determine the temperature of crystallization by homogenization of the fluid inclusions. They concluded that the crystallization of the wyomingite started with the formation of phlogopite and diopside at 1270 °C. Diopside ceased to crystallize at 1220 °C, and leucite appeared abundantly between 1250° and 1150 °C. The solidus temperature was found to be about 1000 °C.

Some heating experiments on natural wyomingite, olivine orendite, orendite, and madupite were made by Fyfe (quoted in Carmichael 1967). The liquidus and the solidus temperatures and the primary phase or phases of the rocks are indicated within brackets after each rock type as follows: wyomingite from Boars Tusk (leucite, 1200° and 1010 °C); wyomingite from Steamboat Springs (leucite, 1165° and 1000 °C); olivine orendite from South Table Mountain (olivine, 1275° and 1010 °C); orendite from North Table Mountain (leucite and olivine, 1215° and 1010 °C); and madupite (leucite, 1245° and 1040 °C). Carmichael (1967) considered that the large melting intervals of these rocks at 1 atm and 1 kb (in the presence of water) suggest that such melting intervals might persist at the base of the crust. If it is assumed that the rocks of the Leucite Hills were produced from an uncontaminated magma, then the large melting intervals noted in these experimental runs were thought by him to disprove their crustal origin. On the basis of these observations, Carmichael concluded that the potassium-rich magma of this area was produced at great depth.

Melting relations of Tristan da Cunha rocks were studied by Tilley et al. (1965). Petrographically these rocks ranged from ankaramite through alkali olivine basalt, trachybasalt to intermediate member of the series, classified as tristanites by Tilley and Muir (1964). Leucite is a modal mineral in the trachybasalt and tristanite (lava of 1961). When the compositions of these rocks are compared with those of Hawaiian and Hebridean series, the Tristan da Cunha rocks are more alkalic (Fig. 15.1), and are particularly rich in K_2O (ranging from 1.47% to 5.03%, Baker et al. 1964). Figure 15.2 shows the trend lines (FeO + Fe_2O_3 versus MgO) of the alkali series of the Hawaii, Hebridean, and Tristan da Cunha provinces. In Fig. 15.2 Tilley et al. plotted the liquidus temperatures of the rock series from these three localities, which shows that the alkali series from Tristan da Cunha displays a less consistent relationship with respect to iron enrichment than the Hawaiian and Hebridean rocks. From their study Tilley et al. concluded that although the rocks of Tristan da Cunha show progressive iron enrichment, accumulation phenomena possibly played a significant role in the consolidation of these rocks, and may be the cause of their having anomalous liquidus temperatures.

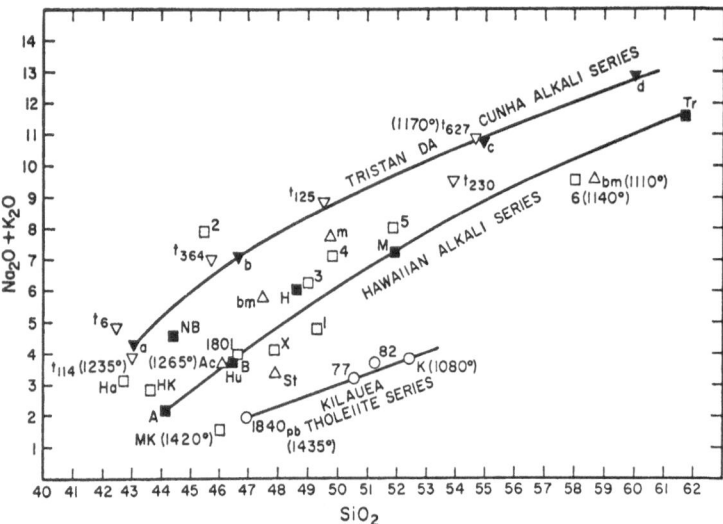

Fig. 15.1. Alkali basalt vs SiO₂ diagram of rocks belonging to Hawaiian, Hebridian, and Tristan da Cunha series. (After Tilley et al. 1965)

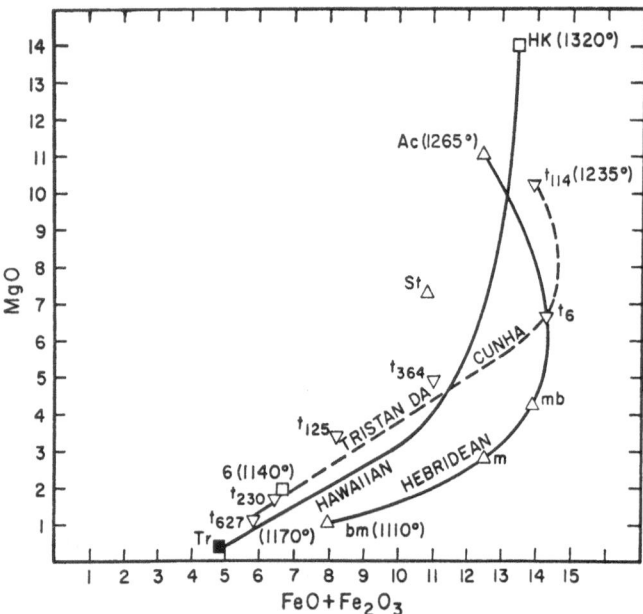

Fig. 15.2. FeO + Fe₂O₃ vs MgO: trend lines of alkali series of Hawaiian, Hebridian, and Tristan da Cunha Province. (After Tilley et al. 1965)

15.2 Studies of Synthetic and Natural Leucite-Bearing Rock Systems Under Different P–T Conditions

Experiments at various pressures on natural leucite-bearing rocks and their synthetic equivalents to determine the nature of their parent materials have so far been few. Some leucite-bearing natural and synthetic rocks were therefore studied under various P–T conditions to understand the genesis of potassium-rich magma in the upper mantle. These results, along with those of other investigators, are described below.

15.2.1 Experiments Under Low Water Pressures (up to 10 kb)

A synthetic melilite-nepheline leucitite ($Di_{28}Ne_{29}Lc_{43}$) was heated at different temperatures (650°–850 °C) and pressures (1 to 5 kb). The starting material was crystallized from a glass of its composition at 900 °C under 1 atmospheric pressure. In this mixture, melilite and olivine were produced by the reaction between nepheline and diopside (Bowen 1922a; Schairer et al. 1962). The results of the experiments are summarized in Fig. 15.3 (bottom part) which shows that melilite disappears at or above 3 kb (AB). Breakdown of leucite occurs at slightly higher pressures (CD, Fig. 15.3). Phlogopite and pyroxene coexist with liquid above DEF, below which phlogopite disappears at low P_{H_2O} by reaction with the liquid, thus supporting the findings of Luth (1967). The reaction, leucite $\rightleftharpoons$ kalsilite + K-feldspar (CE, Fig. 15.3), has been studied by Scarfe et al. (1966; see Fig. 14.1). The results of this investigation suggest that leucite and melilite-bearing rocks are strictly confined to volcanic and subvolcanic conditions.

Experiments in the presence of water vapor ($P_{H_2O} = P_{Total}$) were made (between 1 and 5 kb) in the present study on two natural leucite-bearing rocks, collected from Kunkskopf and Hochsimmer (East Eifel, West Germany, Table 5.13). The rock from Kunkskopf contained phlogopite, plagioclase, nepheline, leucite, apatite, and magnetite in the groundmass with clinopyroxene, and small amounts of olivine as phenocrysts, thus approaching basanitic composition. The rock from Hochsimmer is a tephrite, consisting of clinopyroxene and a small amount of phlogopite as phenocrysts with rare amounts of nepheline, plagioclase, and opaque minerals in the groundmass. At a pressure of 5 kb and temperatures of 800° and 850 °C, the powdered samples of

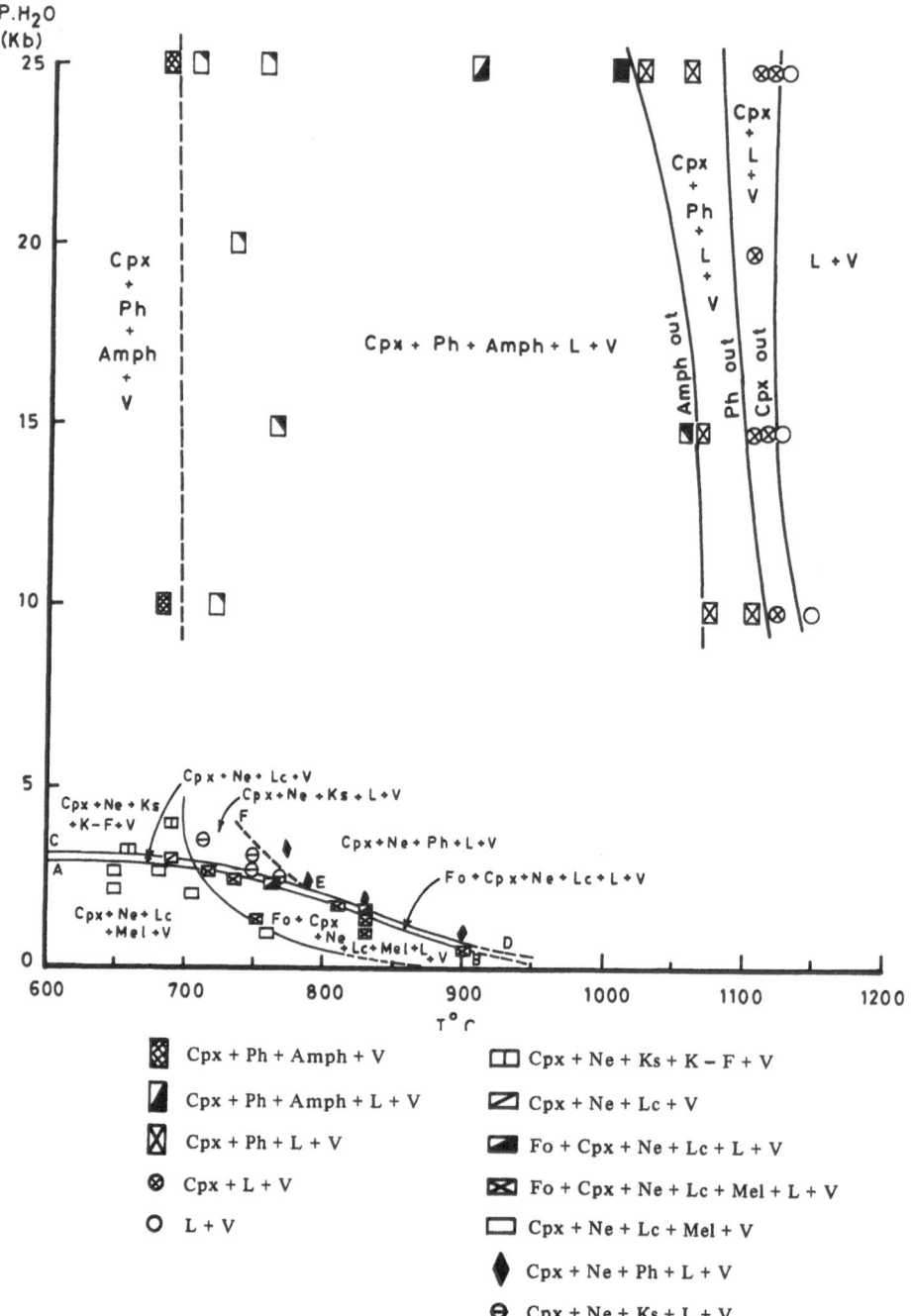

Fig. 15.3. P_{H_2O}–T diagram of a synthetic melilite-nepheline leucitite ($Di_{28}Ne_{29}$-Lc_{43}). (After Gupta et al. 1976). See also List of Abbreviations

the rocks from both areas produced the same assemblage of clinopyroxene, amphibole, phlogopite, magnetite, and liquid, although the
relative proportions of amphibole and phlogopite are less in the Hochsimmer sample. At the same pressure, below 775 °C, garnet and
nepheline appeared in addition to the above assemblage in the Kunkskopf sample. At 5 kb below 750 °C the tephrite from Hochsimmer
produced an assemblage of phlogopite and clinopyroxene with minor
amounts of amphibole and magnetite. Disappearance of leucite was
noted in both cases at and above 1 kb from 600° to 800 °C. At 10 kb
below 800 °C the leucite-nepheline tephrite is represented by an assemblage which corresponds to a magnetite-nepheline-amphibole-
phlogopite pyroxenite. Nepheline disappears above 800 °C and the resulting assemblage at 10 kb is equivalent to a magnetite-amphibole-
phlogopite pyroxenite. Gupta et al. (1976) established that the amphibole is a potassium-richterite.

Yoder (1973) studied an assemblage consisting of akermanite, leucite, and forsterite under fluid pressures of H_2O and CO_2 (the fluid
phases are mixed in different proportions, Chap. 16.3) up to a pressure
of 10 kb. He found that at high pressures the assemblage consisted of
phlogopite, olivine, and calcite, which is similar to the assemblage of
kimberlitic groundmass.

15.2.2 Thermodynamic Considerations for the Conversion
of a Katungite to a Phlogopite Pyroxenite at High Pressures

A synthetic katungite ($Fo_{20}Ak_{40}Lc_{40}$) consisting of forsterite, akermanite, and leucite was studied between 10 and 25 kb P_{H_2O} below
1100 °C and was found to convert to an assemblage consisting of
phlogopite, pyroxene, and liquid. At these pressures liquid appeared at
$750° \pm 25$ °C. Breakdown of diopside$_{ss}$ + phlogopite$_{ss}$ at low pressures
can be explained by the following reaction, if the small amounts of solid
solutions in these phases are ignored:

$$6 \, KMg_3AlSi_3O_{10}(OH)_2 + 4 \, CaMgSi_2O_6 \rightarrow 10 \, Mg_2SiO_4$$
(phlogopite)   (diopside)   (forsterite)

$$+ 2 \, Ca_2MgSi_2O_7 + 6 \, KAlSi_2O_6 + 6 \, H_2O \, .$$
(akermanite)   (leucite)

Table 15.1. Thermodynamic properties of forsterite, diopside, akermanite, leucite, and phlogopite

Phase	$S_{formation}$ (298 K, 1 bar) (cal/deg-gfw)	Volume (cal/bar)	$G_{formation}$ (298 K, 1 bar) (cal/gfw)
Akermanite	−158.66	2.2182	− 879,353
Leucite	−134.188	2.114	− 681,642
Forsterite	− 95.362	1.0466	− 491,983
Diopside	−139.548	1.5795	− 725,784
Phlogopite	−307.964	3.583	−1,389,500

Except for the data of phlogopite, thermodynamic properties of all other phases were either calculated or obtained from Robie and Walbaum (1968). The data of phlogopite was obtained from Bird and Anderson (1973)

Thermodynamic data are available for all phases involved in this reaction (Table 15.1). ΔS_f^0 values for all the phases are calculated from the data of Robie and Waldbaum (1968). The fugacities of water up to 10 kb are given by Burnham et al. (1969). Thermodynamic treatment of the chemical equilibrium involving hydrous and anhydrous phases has been discussed by Fisher and Zen (1971). Using their treatment and the thermodynamic properties of various phases involved in the above reaction (Table 15.1), three sets of equilibrium pressures (P_E) and temperatures (T_E) were determined, which are as follows: (1) P_E = 1415 bar, and T_E = 925 °C, (2) P_E = 1850 bar and T_E = 950 °C, and (3) P_E = 2430 bar and T_E = 980 °C. In this calculation the following considerations were made: (1) the phases involved in the reaction are not solid solutions, (2) the standard state is considered to be 298 °K and 1 bar, and (3) ΔC_P of the reaction is zero. It should be pointed out that a liquid phase may be present in the theoretically determined P_E–T_E conditions and the calculated univariant curve for this reaction would then be metastable. However, the data, when used in a qualitative sense, do indicate that an assemblage consisting of forsterite, akermanite, and leucite should be stable only at high temperatures and low P_{H_2O}, but would be converted to a phlogopite and pyroxene-bearing assemblage at high water pressures. The above theoretical study therefore suggests that katungites have been formed under volcanic or subvolcanic conditions, and they are represented at greater depth by phlogopite pyroxenites.

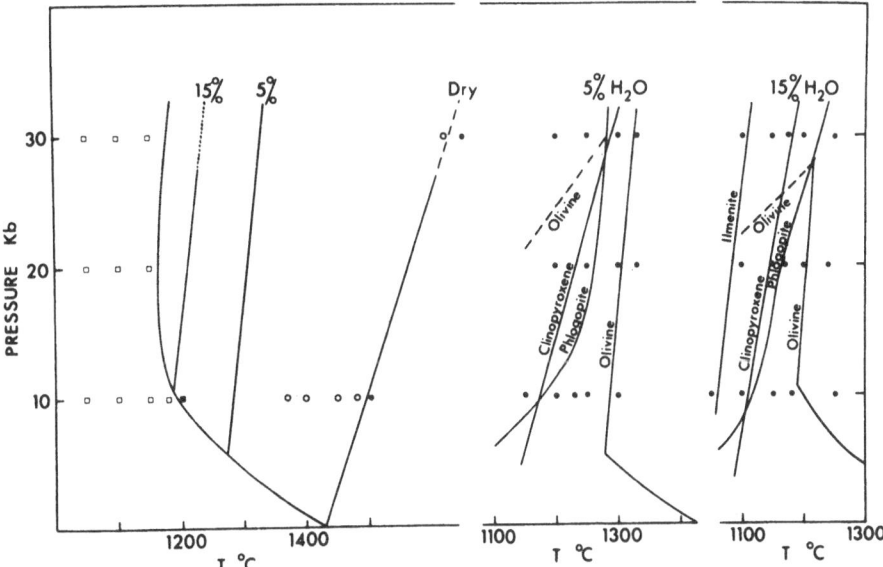

Fig. 15.4. P_{H_2O}–T diagram of a synthetic biotite mafurite. The *left hand diagram* shows subsolidus *(open symbols)* and above-solidus runs for water-saturated and anhydrous conditions and the positions of the liquid for 5% and 15% H_2O. Olivine is the liquidus phase along liquidi, shown as *solid lines*. The *center diagram* shows experimental points and incoming of phases for 5% H_2O and the *right diagram* shows similar data for 15% H_2O. The olivine-out boundary at high pressure, due to reaction with liquid and other solid phases with decreasing temperature, is shown as a *dashed boundary*. (After Edgar et al. 1976)

15.2.3 Study of a Synthetic Biotite Mafurite Glass at High Pressures

A synthetic biotite mafurite glass was studied by Edgar et al. (1976) between 10 and 30 kb under dry and wet conditions. A natural biotite mafurite of similar composition, studied by Holmes (1942), consisted of olivine, kalsilite, perovskite, and biotite with xenoliths of glimmerite. Edgar et al. noted that the liquidus ranged from 1490 °C at 10 kb to 1625 °C at 30 kb (Fig. 15.4), clinopyroxene being the primary phase at higher pressures and olivine at lower pressures.

The liquidus temperatures, however, were lowered by 200°–300 °C, in presence of 5% water, olivine being the primary phase. Their runs with 15% water, show a further lowering of temperatures, olivine still remaining as the liquidus phase up to a pressure of slightly lower than 30 kb; above which phlogopite became the primary phase.

15.2.4 Study of a Synthetic Leucite Basanite
and a Melilite-Nepheline Leucitite Between 10 and 25 kb

The mafurite studied by Edgar et al. (1976) contained 7.1 wt. % K_2O, and the study is important for the genesis of potassium-rich magma. However, a biotite mafurite is not a representative sample of leucite-bearing rocks. Leucite tephrites, basanites, and melilite-nepheline leucitites are more commonly occurring rock types in highly potassic petrographic provinces (Chap. 5), and further varieties of rocks containing leucite can be produced by fractionation of magmas of these compositions. A leucite basanite ($Fo_{20}Di_{30}Lc_{30}An_{20}$) and a melilite-nepheline leucitite were therefore studied at high water pressures,

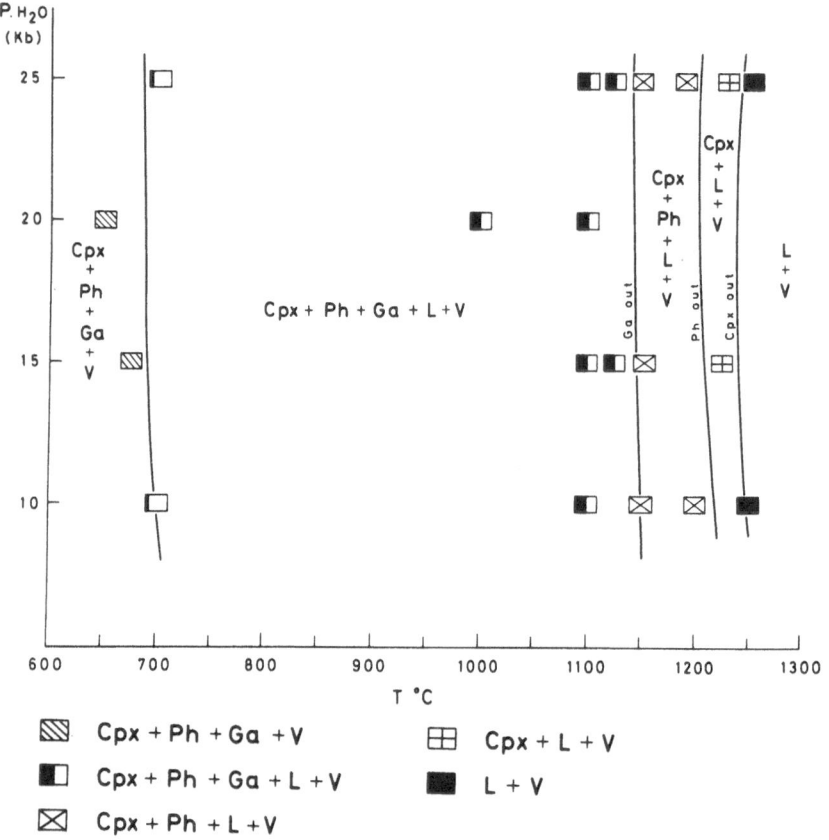

Fig. 15.5. P_{H_2O}–T diagram of a synthetic leucite basanite ($Fo_{20}Di_{30}Lc_{30}An_{20}$). (After Gupta et al. 1976). See also List of Abbreviations

equivalent to upper mantle conditions. The compositions were chosen in such a way that the proportion of different phases was close to their natural equivalents.

Results of these experiments are summarized in Figs. 15.3 and 15.5. Determination of cell parameters by Gupta et al. (1976) indicated that all phases are solid solutions. Figure 15.5 shows that the leucite basanite is represented at low temperatures between 10 and 25 kb by $garnet_{ss}$, $phlogopite_{ss}$, and $diopside_{ss}$. Clinopyroxene is the primary phase, whereas garnet is the last phase to appear.

With the mixture $Di_{28}Ne_{29}Lc_{43}$, leucite was the primary phase at 1 atm. The appearance of leucite at 1197 °C (in air) was followed by $forsterite_{ss}$ (1194 °C), $nepheline_{ss}$ (1185 °C), melilite (1178 °C), and $diopside_{ss}$ (1100 °C); $forsterite_{ss}$ disappeared at 1160 ± 10 °C. The composition of melilites and $nepheline_{ss}$, determined by using the methods of Edgar (1965) and Hamilton and MacKenzie (1960) were respectively as follows: $Ak_{70}Sm_{30}$ and $Ne_{68}Ks_{29.8}Si_{2.2}$. Figure 15.3 shows that melilite-nepheline leucitite and olivine-melilite-nepheline leucitite are represented at low temperatures and high pressures (between 10 and 25 kb) by amphibole and phlogopite-bearing pyroxenites. Clinopyroxene is the primary phase again as in the case of basanite, although it was the last phase to crystallize at 1 atm. The results of the investigation show that the liquidus temperature of 1197 °C is lowered by 82° and 64 ± 5 °C at 25 and 10 kb respectively. The cell parameter and the microprobe data suggested that the amphibole is a $richterite_{ss}$. Mysen and Boettcher (1975; p. 569) studied natural periodites with or without additional synthetic phlogopite. They found that the compositional range of amphiboles produced at high pressures was between magnesian paragasitic hornblende and tschermakitic hornblende. However, when the same peridotites were spiked with synthetic phlogopite, the amphiboles produced became considerably potassic.

Experiments were made on the synthetic melilite-nepheline leucitite $(Di_{28}Ne_{29}Lc_{43})$ at 750 °C, 20 and 25 kb under dry conditions, which produced an assemblage consisting of omphacite and kalsilite. The formation of such an assemblage can be explained by the following reaction:

$$CaMgSi_2O_6 + NaAlSiO_4 + KAlSi_2O_6 \rightarrow CaMgSi_2O_6 \cdot NaAlSi_2O_6$$
(diopside)   (nepheline) (leucite)   (omphacite)

$$+ KAlSiO_4 .$$
(kalsilite)

In contrast to the composition of pyroxene produced under dry conditions, extensive partitioning of sodium into pyroxene did not take place in the experiment under high water vapor pressures. Instead, sodium was contained in the amphibole. However, if the starting material ($Di_{28}Ne_{29}Lc_{43}$) contained additional small amounts of forsterite, under a condition of $P_{Total} \gg P_{H_2O}$, a phlogopite pyroxenite (the pyroxene being omphacite) should be produced by the following reaction:

$CaMgSi_2O_6$ + $NaAlSiO_4$ + 2 $KAlSi_2O_6$ + 3 Mg_2SiO_4
(diopside)   (nepheline) (leucite)   (forsterite)

+ 2 H_2O → $CaMgSi_2O_6$, $NaAlSi_2O_6$
(omphacite)

+ 2 $KMg_3AlSi_3O_{10}(OH)_2$.
(phlogopite)

Thus a melilite-nepheline leucitite and a phlogopite pyroxenite (omphacite constituting the pyroxene component) are isochemical. To produce plagioclase-bearing basanites and tephrites, the phlogopite pyroxenite should contain garnet in addition. In this connection mention may be made of the work of Kushiro and Erlank (1970), who studied the stability of potassium richterite in the presence of garnet ($CaMg_2Al_2Si_3O_{12}$), in the ratio 1:1 (also see Chap. 14.4). Experiments up to 30 kb in the presence of excess water showed that richterite and garnet reacted to form phlogopite and clinopyroxene with enstatite (part or all of which may be dissolved in solid solution). The reaction suggested by them was as follows:

$KNaCaMg_5Si_8O_{22}(OH)_2$ + $CaMg_2Al_2Si_3O_{12}$ → $KMg_3AlSi_3O_{10}(OH)_2$
(richterite)   (garnet)   (phlogopite)

+ (2 $CaMgSi_2O_6$) · ($NaAlSi_2O_6$)
(omphacite)

+ 2 $MgSiO_3$.
(enstatite)

Further implications of the results of investigations on leucite-bearing assemblages under different pressures will be described in Chap. 18.

Chapter 16 Leucite-Bearing Rocks and Their Relation to Kimberlites

16.1 Comparison of the Chemistry of Kimberlites and Leucite-Bearing Rocks

Chemical affinity between leucite-bearing rocks and kimberlites has been pointed out by several workers (Holmes 1932; Wade and Prider 1940; Borley 1967; Cundari 1973). Some petrologists such as Holmes (1932), O'Hara (1965), O'Hara and Yoder (1967), and Yoder (1973) considered that these two groups of rocks are genetically related. Because of this, the chemistry of these two groups of rocks is compared below. Possible genetic connection between kimberlites and leucite-bearing rocks is discussed in the following section.

Chemical composition of 10 kimberlites from various localities is given in Table 16.1, whereas that of leucite-bearing rocks is presented in Tables 5.1 to 5.10 and 5.13 to 5.15. Comparison of these tables shows that the average MgO content of kimberlites (25%) is much higher than that of the leucitic rocks (7%). In many Italian rocks MgO content may be lower than even 5%. In some rare cases such as kalsilite katungite from Bunyarugura, Uganda (13.54%), jumillite from southern Spain (16.88%), and olivine leucitite from New South Wales (13.51%) the MgO content is higher compared to most potassic rocks but even these figures are much lower than the magnesia content of kimberlites.

While there are exceptions, the SiO_2 content of most leucite-bearing rocks lies between 44% and 52%, whereas the silica content of kimberlites ranges between 30% and 35%. Although in some rare instances the SiO_2 content of potassic rocks is low, such as in some kalsilite katungites from Bunyaruguru (33.52%) or wolgidites from West Kimberley (36.02%), most leucite-bearing rocks from Italy have much higher silica content, often exceeding 50%. The potassic rocks with very low silica content (mentioned above) are much poorer in MgO and richer in CaO and thus do not resemble kimberlitic composition.

The average Al_2O_3 content of kimberlites is around 5% but in case of leucite-bearing rocks the mean of 89 analyses from various localities

Table 16.1. Chemical compositions of kimberlites from various localities

	1	2	3	4	5	6	7	8	9	10
SiO_2	31.60	35.47	32.23	42.26	27.98	33.21	31.1	35.6	36.12	28.52
TiO_2	2.02	1.71	1.48	2.47	4.22	1.97	2.03	3.06	1.46	2.68
Al_2O_3	3.21	6.84	1.58	5.14	2.64	4.45	4.9	3.46	4.38	5.95
Fe_2O_3	6.33	10.18	5.49	5.56	8.15	6.78	10.5	4.60	6.80	7.40
FeO	3.37	1.24	3.42	3.32	4.25	3.43		8.71	2.68	4.02
MnO	0.30	0.18	0.15	0.14	0.18	0.17	0.10	0.20	0.22	0.22
MgO	29.45	16.52	31.88	17.49	26.17	22.78	23.90	27.90	22.82	20.36
CaO	8.07	5.95	6.14	7.50	9.16	9.36	10.60	6.78	8.33	14.72
Na_2O	0.16	0.18	0.04	0.25	0.64	0.19	0.31	0.82	0.29	0.06
K_2O	0.34	0.28	0.97	0.67	1.78	0.79	2.1	2.00	5.04	1.90
H_2O^+	11.37	8.41	8.88	8.88	7.33	2.66	5.9	3.72	4.89	7.67
H_2O^-	0.87	7.68	0.83	0.83	0.40	8.04			1.28	0.24
P_2O_5	0.94	0.56	0.29	0.51	0.94	0.65	0.66	0.40	1.46	1.21
Cr_2O_3		0.13	0.23	0.08	0.14	0.54				0.15
CO_2	1.96	0.16	5.11	5.11	5.83	4.58	7.1	2.42	3.80	3.74
S_2^-		0.02	0.12	0.13	0.09					0.88
Total	99.99	95.51	98.84	100.34	99.90	99.60	99.20	99.67	99.57	99.72
Less O = S		0.01	0.06	0.06	0.04					0.44
Total	99.99	95.50	98.78	100.28	99.86	99.60	99.20	99.67	99.57	99.28

1. Kimberlite from Kao pipe, Basutoland, S. Africa. Anal. M.H. Kerr (Nixon and Boyd 1973a)
2. Kimberlite from the Lemphane area, Lesotho, S. Africa. Anal. Gurney and Ebrahim (Kresten and Dempster 1973)
3. Kimberlite from the same area as No. 2
4. Kimberlite from the Sekameng area. Anal. Gurney and Ebrahim (Dempster and Tucker 1973)
5. Kimberlite from Monastery Mine, Lesotho, S. Africa. Anal. F.R. Boyd (Nixon and Boyd 1973)
6. Average analyses of kimberlites (Gurney and Ebrahim 1973)
7. Micaeous kimberlite (Dawson 1968)
8. Kimberlite from Saglek, Labrador (Collerson and Malpas 1977)
9. Kimberlite from New Elands Mine, S. Africa (Dawson and Smith 1977)
10. Kimberlite from Hard Bank, Lesotho, S. Africa (Gurney and Ebrahim 1973)

is 14%. In case of many Italian leucitic rocks alumina content is as high as 18% to 20%.

The K_2O/Na_2O ratio of both kimberlites and leucite-bearing rocks is always > 1, but the Na_2O content of the former is very low, average being around 0.3%, whereas the average soda content of 89 leucite-bearing rocks is 2.6%. Although jumillites from southern Spain and some potassic rocks from the Leucite Hills, West Kimberley, and New South Wales have low Na_2O ($\simeq 1\%$), in other areas its concentration ranges between 3% to 5% and consequently nepheline becomes an important mineral.

Although in some micaceous kimberlites, the K_2O content can be as high as 5% (Table 16.1; Dawson and Smith 1977) the average potash content of kimberlites is usually $\simeq 1$, much lower than the K_2O content of leucite-bearing rocks (5%). As discussed in Chap. 1.1, the potash content of the rocks from the Leucite Hills and West Kimberley is as high as 9% or 10%. In some rare instances the potash content of some rocks from Villa Senni, Italy, and Celebes, Indonesia may be as high as 17% and 19.98% respectively.

The lime content of leucitic rocks is quite variable, the average being 8%, which is comparable to kimberlites. In phonolites and trachytes, pyroxene and plagioclase contents are very low, thus giving a low CaO value. However, the tephrites and basanites from the East Eifel (Table 5.14; 11.8% to 12.8%), leucitites from Kitale, Uganda (12.72%, leucitites from Roccamonfina (11.6%) and wolgidites from West Kimberley (15.12%) have high CaO contents.

The average Σ FeO* content of kimberlites (10%) is comparable to that of leucite-bearing rocks (8%). However, like kimberlites, potassic rocks of various areas display a wide range of total iron oxide content. For example the average Σ FeO content of the potassic rocks from southern Spain, Utsuryo Island, Italy, and the Leucite Hills is 5.5% ± 0.5%, whereas the rocks from New South Wales, Uganda, Manchuria, and the East Eifel have an average of 11%. The average Σ FeO content of other areas lies in between.

The average TiO_2 content of both kimberlites (2.3%) and leucite-bearing rocks (2.1%) is quite similar. However, potassic rocks from some localities such as Java and Celebes and West Kimberley have TiO_2 content as high as 5% to 7%.

---

* Σ FeO refers to $FeO + Fe_2O_3$

The above discussions suggest that the MgO, Al_2O_3, and Na_2O contents of both groups of rocks are significantly different, and although both kimberlites and leucitic rocks are SiO_2-undersaturated, potassic rocks are much more silicic in character. Kimberlites in general are only very slightly potassic, whereas leucite-bearing rocks are highly potassic. However, phlogopite-bearing kimberlites have high K_2O content and this variety has some affinity to leucitic rocks, but even they are more enriched in MgO and very poor in Al_2O_3 and Na_2O.

Both groups of rocks are rich in Zr, P, Ba, Sr, and F (Wade and Prider 1940).

16.2 Possible Genetic Connection Between Potassic Rocks and Kimberlites

Holmes (1932) suggested that separation of eclogite or dunite from a peridotite magma under high volatile pressure would produce kimberlites; if, however, volatiles escape, the resulting rocks would be similar to olivine leucitite. Eclogites are usually high in their Na_2O/K_2O ratio and relatively low in their silica content, whereas peridotites are poorer in their sodium content. Extraction of eclogite may therefore generate a potassium-rich residual liquid. Problems related to this hypothesis will be discussed in Chap. 18.

16.3 Experimental Study of a Synthetic Katungite Under P_{H_2O} and P_{CO_2} in Relation to Kimberlite Genesis

The genetic relationship between kimberlites and olivine melilitites was stressed by Lewis (1897), Wagner (1914), Shand (1934), Taljaard (1936), and Holmes (1937). More recently Ukhanov (1963) found association of kimberlite breccia and olivine melilitites in Bargydamalakh pipe (Anabar region, U.S.S.R.), where olivine melilitites form the core of kimberlite breccia. Olivine and melilite are often associated with leucite, kalsilite, and phlogopite in leucite-bearing rocks. Yoder (1973) discussed the relationship between melilite and leucite-bearing rocks with kimberlites. He considered that a simplified katungite may be related to an olivine-kalsilite pyroxenite (mafurite) by the following reaction:

$2 Ca_2MgSi_2O_7 + 3 KAlSi_2O_6 + (X + 1) Mg_2SiO_4 \rightarrow$
(akermanite)   (leucite)   (forsterite)   (16.1)

$4 CaMgSi_2O_6 + 3 KAlSiO_4 + X Mg_2SiO_4$.
(diopside)   (kalsilite)   (forsterite)

He found that addition of H_2O and CO_2 in different proportion to the product of Eq. (16.1), resulted in an assemblage, which is similar to kimberlite groundmass. The assemblage consisting of phlogopite, calcite, and forsterite can be explained by the following reaction (Yoder 1973):

$CaMgSi_2O_6 + 3 KAlSiO_4 + (X + 4) Mg_2SiO_4 + H_2O$
(diopside)   (kalsilite)   (forsterite)

$\qquad + CO_2 \rightarrow 3 KMg_3AlSi_3O_{10}(OH)_2 + CaCO_3$ (16.2)
$\qquad\qquad$ (phlogopite)   (calcite)

$\qquad + X Mg_2SiO_4$.
$\qquad$ (forsterite)

From his study Yoder concluded that "there is some support that magma having melilite affinities could be transformed into kimberlite with addition of suitable volatiles. Loss of volatile in transit or crystallization at low pressure would yield a melilite-bearing assemblage within the upper crust".

16.4 Experimental Study of Two Picrites with Reference to the Genesis of Kimberlite and Potassium-Rich Magma

O'Hara and Yoder (1967) considered that separation of eclogite from a picritic magma at depth (80–125 km) may produce a liquid with high K_2O/Na_2O ratio. Carmichael (1967) also considered that such a process should produce a liquid not only high in the K_2O/Na_2O ratio but also in Rb, Cr, and Ni as found in these two groups of rocks. A nepheline-normative picritic diabase and a hypersthene-normative picritic basalt were therefore studied by Gupta and Yagi (1977) at temperatures and pressures up to 1300 °C and 30 kb in presence of excess water. Chemical analyses of these two rocks and their normative compositions are given in Table 16.2. Results of experiments on these

Table 16.2. Chemical and normative compositions of picritic basalts, used for experimental investigation

	1	2			1a	2a
SiO_2	46.07	47.47	Or		11.68	3.56
TiO_2	1.98	0.71	Ab		12.58	12.73
Al_2O_3	11.06	17.09	An		12.51	38.03
Fe_2O_3	5.00	1.32	Ne		5.11	–
FeO	5.44	7.31	Di	Wo	18.33	2.62
MnO	0.20	0.24		En	14.50	1.71
MgO	9.97	11.31		Fs	1.72	0.73
CaO	11.73	9.15	Hy	En	–	17.20
Na_2O	2.59	1.51		Fs	–	7.39
K_2O	2.01	0.60	Ol	Fo	7.28	6.68
H_2O^+	2.22	2.56		Fa	1.02	2.71
H_2O^-	1.78	0.64	Mt		7.19	1.86
P_2O_5	0.23	0.18	Il		3.80	1.37
Total	100.28	100.09	Ap		0.67	0.40

1, 1a. Picritic dolerite No. 3001 from Nosappu Cape. Anal. N. Onuki (Yagi 1969)
2, 2a. Picritic basalt No. 65851 from Wakuike. Anal. H. Matsumoto (Takeshita 1974, 1975)

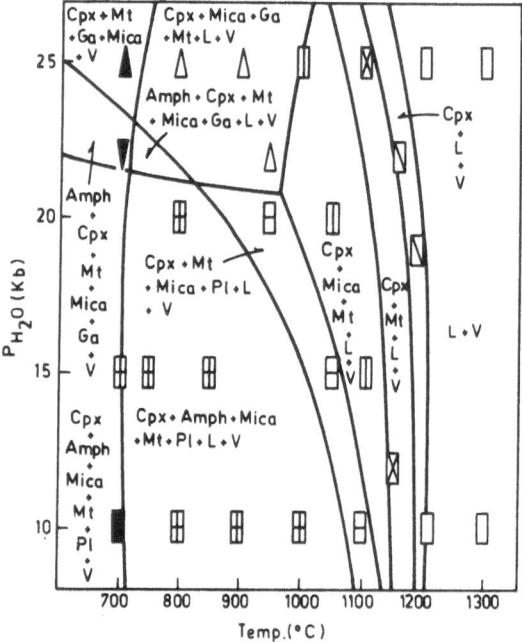

Fig. 16.1. P_{H_2O}–T diagram of a picritic dolerite from Nosappu Cape. (After Gupta and Yagi (1977). See also List of Abbreviations

Fig. 16.2. P_{H_2O}–T diagram of a picritic basalt from Wakuike. (After Gupta and Yagi (1977). See also List of Abbreviations

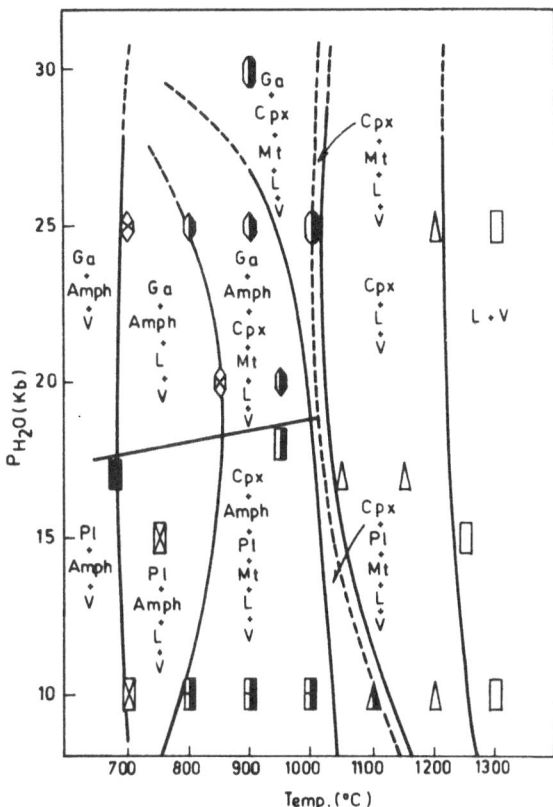

two rocks are summarized in Figs. 16.1 and 16.2. In case of the nepheline-normative picrite (Fig. 16.1), the composition of a liquid formed at 800 °C and 25 kb (Table 16.3) indicates a decrease in SiO_2 and CaO and increase in MgO, FeO, and TiO_2 contents, compared to the original rock (Table 16.2). However, the CaO, Al_2O_3, and Na_2O contents are still much higher compared to average kimberlites from South Africa (Gurney and Ebrahim 1973). The analyzed liquid coexists with slightly sodic pyroxene and garnet (for analyses see Table 16.3) and thus corresponds to a picritic liquid, which has lost its eclogitic fraction. The Na_2O/K_2O ratio of this liquid is, however, > 1, thus contradicting the suggestion of O'Hara and Yoder (1967). The composition of the liquid described here, is however, somewhat similar to nephelinites, which may support the proposal of Green and Ringwood (1967) that subtraction of eclogitic material from a picritic magma may produce a magma of nephelinitic composition.

Table 16.3. Analyses of crystalline phases and liquids

	1	2	3	4	5	6	7
SiO_2	37.57	49.43	39.23	39.57	49.31	46.76	46.67
TiO_2	–	–	3.60	–	0.49	0.27	0.35
Al_2O_3	19.79	3.52	13.73	20.41	1.94	9.06	12.26
FeO	21.44	8.26	12.51	21.71	13.58	15.55	13.76
MnO	0.41	0.15	0.18	0.29	0.44	0.50	0.43
MgO	8.95	14.09	13.41	10.42	14.68	14.53	15.40
CaO	11.85	22.00	8.87	9.41	16.81	8.32	6.14
Na_2O	0.08	2.82	2.80	0.08	0.35	2.05	0.58
K_2O	–	0.06	1.74	–	0.44	0.32	0.10
Cr_2O_3	0.21	0.06	–	0.15	0.02	–	–
Total	100.30	100.39	96.07	102.04	98.06	97.36	95.69

1. Garnet, crystallized at 800 °C and 25 kb from Nosappu picritic dolerite
2. Clinopyroxene, crystallized at 800 °C and 25 kb from Nosappu picritic dolerite
3. Liquid, formed at 800 °C and 25 kb from Nosappu picritic dolerite
4. Garnet, crystallized at 1000 °C and 25 kb from Wakuike picritic basalt
5. Clinopyroxene, crystallized at 1000 °C and 25 kb from Wakuike picritic basalt
6. Amphibole, crystallized at 800 °C and 20 kb from Wakuike picritic basalt
7. Liquid, formed at 1000 °C and 25 kb from Wakuike picritic basalt
In analyses Nos. 3, 6, and 7 the deficiency from 100 corresponds to the volatile components

In case of the hypersthene-normative picrite, the liquid produced at 1000 °C and 25 kb is in equilibrium with garnet, clinopyroxene, and magnetite and has SiO_2 (46.67%) and MgO (15.40%) contents similar to that of some picritic basalts rather than kimberlites or liquids of high K_2O/Na_2O ratio.

Experimental investigation of Gupta and Yagi (1977) on both picrites therefore shows that at a pressure of 25 kb, subtraction of eclogitic fraction from a nepheline-normative or a hypersthene-normative picritic basalt does not produce liquids with K_2O/Na_2O ratio > 1.

Chapter 17 Structural and Tectonic Control of Alkali Magmatism with Special Reference to Leucite-Bearing Lavas

17.1 Volcanism Associated with Rift Zones and Fault Systems

Major alkaline and peralkaline igneous provinces are found to be confined to shield and stable continental areas of the crust (Bailey 1974), which was described by Bucklund (1933) as the epirodiatresis relationship. Thus a special tectonic control may be necessary to generate such magmas in these regions. Bailey studied the tectonic control of alkali magmatism in East and Central Africa. In this region the localization of volcanic activities (producing alkaline and carbonatite rocks) has been noted along Rift Valleys by Holmes and Harwood (1932, 1937) and King and Sutherland (1966). Bailey (1974) described the rift zones as long belts of structural activities, characterized by the presence of various combinations of fault troughs, faulted monoclines, block faulting, and transverse faults. Alkali volcanism related to such a rift zone has also been described from the West African province (Black and Girod 1968, 1970), Western USA (Erdley 1961), the Midland Valley of Scotland, the French Massif Central, and the Laacher See area of West Germany (Ahrens 1961). Bailey (1964) found that the rifts are actually broken crests of crustal arches, which have been uplifted above surrounding broad basinal areas. He thought that the uplifting took place in several distinct stages, which were marked by erosional surfaces. According to him preservation of these uplifted erosional surfaces would need structurally competent crust. Bailey also noted that most intense magmatism was also associated with greatest uplift. Harris (1969) suggested that heating of the underlying mantle would cause its expansion, which would result in the uplift of the overlying crust. Both uplift and associated volcanism are thus related to the thermal state of the underlying mantle below the rift zones. Gass (1970) also proposed a similar mechanism. However, the rifts and associated volcanism date back to Cretaceous and perhaps Precambrian times, and therefore these fractures have been present for a long geologic time. If the rift systems were related to thermal expansion of the underlying mantle, these uplifts would have collapsed during the dissipation of heat. The sources of heat are also found to be concentrated along long narrow belts. Bailey (1974) considered that the mantle heating could not be related to mantle convection, as this would have produced crustal separation during this long period of existence of the rift system in East and Central Africa. He also thought that this mechanism would require a complex pattern of convection cells. He considered that the basin and swell pattern of the African continent might be the result of a radial compressive force on a rigid continental plate. According to him such a force might have originated in the girdle of mid-oceanic ridges. If an arch is produced in a rigid continental plate, partial melting would result due to decompression in the underlying zone of reduced pressure. This would further cause the flow of volatiles in the zones of pressure relief, which would also bring additional heat and reduce the melting ranges of the rocks therein.

Wade and Prider (1940) found that a great number of the outcrops of leucite-bearing rocks in the Western Kimberley province of Australia are found along synclinal depression between the Mt. Wynne anticline and anticlinal folds of the St. George's Range. They found that the rocks occur along two directions: (a) north-northeast to south-south-west, (b) west-northwest to east-southeast. These two directions are important fault and structural lines of western Australia in general. The distribution and the structure of the volcanic masses show that the magma ascended along the fault planes, linked with the structure of the Precambrian basement rocks.

The leucitite lavas of the New South Wales region of Australia were recently studied by Cundari (1973). He considered the lava of that region to be homogeneous and derived from great depths. The eruption of the lava through various independent vents would require favorable condition for their transit to the surface, which he thought to have been generally uninterrupted especially in the low pressure region. Cundari suggested that the deep-seated crustal structure of the basement complex, reactivated in the late Cenozoic period by faulting and differential vertical movements, helped the ascent of the leucititic magma to the surface. His argument is thus consistent with Bailey's (1964) idea.

17.2 Tectonics in the East Eifel Area of W. Germany

Ahrens (1961) described the structure and tectonic history and their relation to volcanic activities in the Laacher See area of West Germany.

According to him volcanism in this area was connected with the development of the Neuwieder basin. Ahrens considered that the structural movements related to the development of the Neuwieder basin started at the beginning of Oligocene. The main movements which took place during the change of Pliocene to Pleistocene were accompanied by extensive lifting and faulting of the Rhine Shield. The eruption of the Laacher See area began immediately after these movements and reached a climax. Volcanism in the Westerwald area began during Pliocene. The points of eruptions of the Quaternary volcanic rocks, tuffs, and gaseous emanations are absent in the eastern part of the Neuwieder basin. There is also no point of eruption beyond the fault which is situated near the Rhine valley between Brohl and Andernach and its southeastern extension. Such distribution of the points of eruption can possibly be explained by the presence of the magma chamber or chambers almost vertically below these points; these points are also places of carbonic acid emanations. Cloos (1939) believed that the magma was poured out through a far-reaching fissure, striking north-northwest–south-southeast, which corresponds to the strike direction of today's Rhine valley. The deep tectonic sinking in the Neuwieder basin forced the magma to the northwest or west as it reached the contact zone between the crystalline basement and its overlying sediments. From the study of the nature of crystalline ejecta he concluded that the magma remained for some time in the crystalline basements. Seismic data are not available to determine the exact depth of contact plane between the folded sequence and the basement rocks.

17.3 Plate Tectonic Model for the Generation of Potassium-Rich Magma

17.3.1 Origin of Highly Potassic Magmas of Indo-Pacific and Calabrian Arcs

Kuno (1959) and Sugimura (1960) showed that alkali contents of volcanic rocks in Japan and surrounding areas increase systematically with increase in depth of earthquake foci underneath the volcanoes.

Hatherton and Dickinson (1968, 1969) plotted K_2O (instead of $K_2O + Na_2O$) versus SiO_2 contents of the volcanic rocks from the Pacific, Indonesian, and Atlantic island arc regions as a function of the earthquake focal depths and showed even better correlation than in the case of the total alkalis. In their study they did not include volcanoes of Batu Tara and Murriah (Indonesia) and Utsuryo island (Japan Sea). These volcanoes are underlain by deep focal earthquakes (> 300 km) and were not considered by them as island arc volcanoes.

Ninkovich and Hays (1972) extended the studies of Hatherton and Dickinson and included Batu Tara, Murriah and various localities of the Calabrian arc in their studies. All these localities are characterized by high K_2O content. In their diagrams of K_2O versus SiO_2, Ninkovich and Hays plotted the curves representing similar earthquake focal depths (Figs. 17.1 and 17.2). From these studies they concluded that Indonesian volcanoes of Batu Tara (north west of Java) lie about 300 to 400 km above the Benioff zone. The rocks of these volcanoes have K_2O contents similar to those of the Mediterranean region (Figs. 17.1 and 17.2). The volcanoes of Utsuryo island and Manchuria (K_2O/SiO_2 ratio as high as those of Batu Tara and Murriah) are considered by them to be about 600 km above the inclined seismic zone. According to Ninkovich and Hays (1972) the active volcanoes of the Mediterranean are related to a deep-seated tectonic process. The epicenters of the earthquakes in the Tyrrhenian Sea region are shown in Fig. 17.3. Shal-

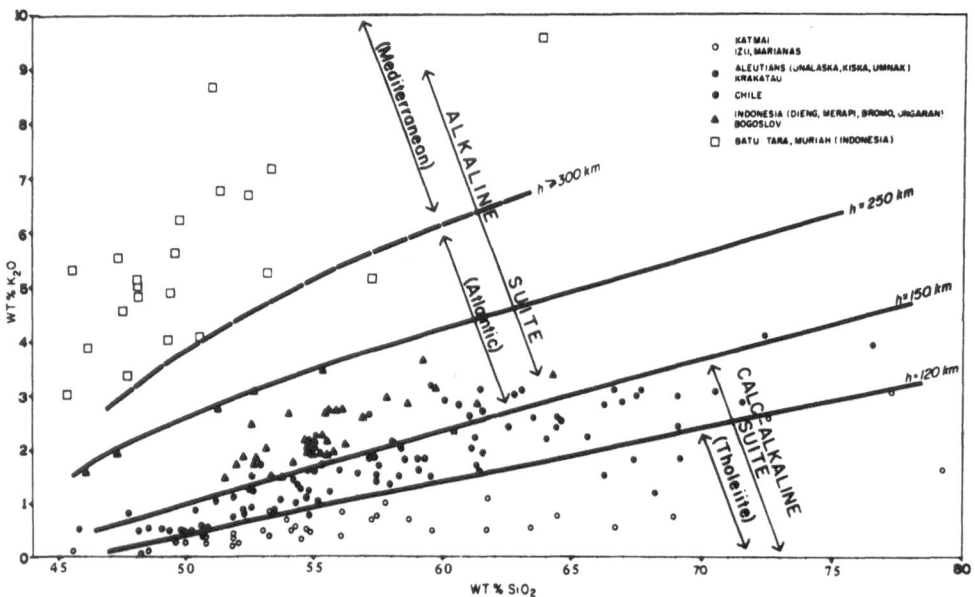

Fig. 17.1. K_2O and SiO_2 contents of volcanic rocks from Indo-Pacific areas. *Curves (h)* refer to the depths of Benioff zone beneath volcanoes. (After Ninkovich and Hays 1972)

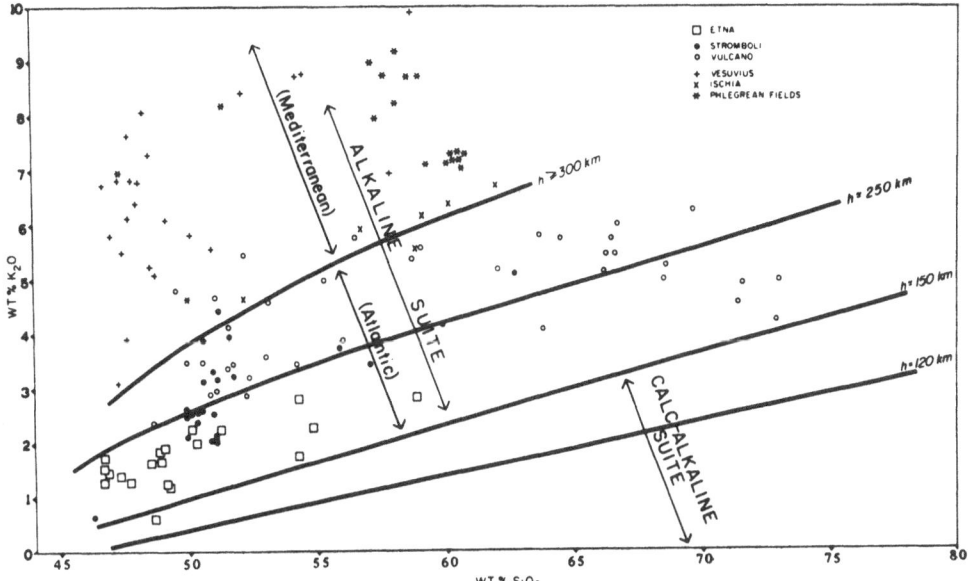

Fig. 17.2. K_2O and SiO_2 contents of volcanic rocks from Calabrian arc and curves of equal earthquake focal depths in the Indo-Pacific region as mentioned in Fig. 17.1. (After Ninkovich and Hays 1972)

low earthquakes are associated with the axis of the orogenic belts, passing through Sicily and the Apennine peninsula. The intermediate earthquakes are distributed along an inclined plane, which deepens from a possible buried trench, south of Sicily and Calabria toward the Neopolitan area, where it reaches a depth of more than 300 km. Earthquake foci with depth between 100 and 200 km are rare in the Tyrrhenian Sea. The distribution of the earthquake foci in the Tyrrehenian Sea was interpreted by Peterschmidt (1957) as evidence of an inclined Benioff zone, dipping under the Calabrian arc (Ninkovich and Hays 1972). According to these workers, if the plane of intermediate depth earthquakes below the Neopolitan area and Eolian arc region is extended south of Italy, then the plane would have a depth of about 100 to 200 km below Etna. However, no intermediate depth earthquakes are recorded under Mount Etna of Sicily.

The volcanism in the Tyrrhenian Sea has been explained by Ninkovich and Hays in terms of plate tectonic model as shown in Figs. 17.4 and 17.5. According to this model, during subduction, as the lithosphere enters the asthenosphere, it is heated, followed by subsequent

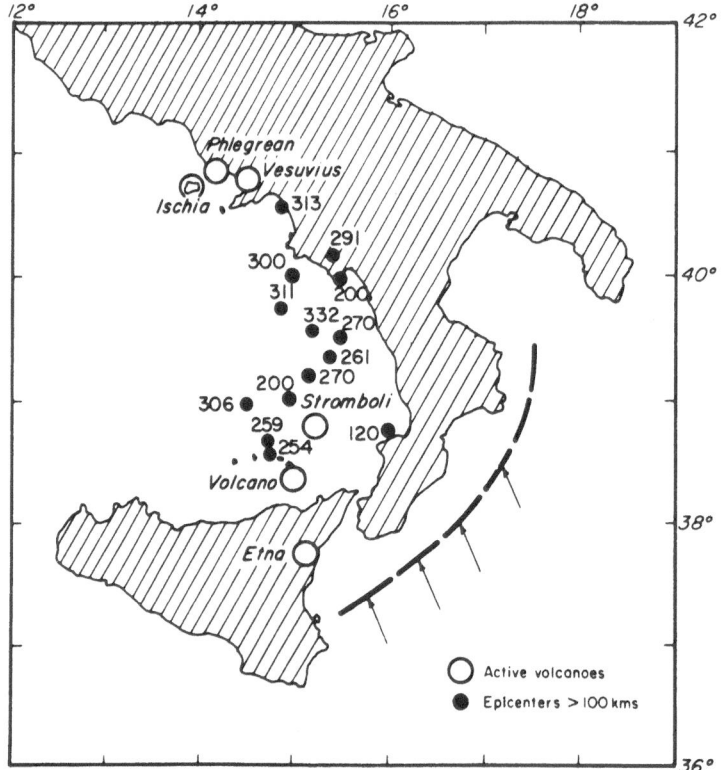

Fig. 17.3. Calabrian arc. Distribution of active volcanoes and earthquakes with focal depths greater than 100 km. (After Ninkovich and Hays 1972)

loss of water. Raleigh (1967) considered that the seismicity of the Benioff zone is the result of a fracture caused by the dehydration of the subducted crustal blocks as they are pulled down by the mantle convection. Sugimura (1960) and McBirney (1969) believed that dehydration of the crustal block is important for the melting of the associated mantle materials. Ninkovich and Hays considered that water would migrate upward, scavenging potassium, rubidium, sodium, and other cations from the surrounding asthenosphere. The amounts of these materials extracted by such a process should be related to the temperature of H_2O and the distance traveled through the asthenosphere. The water thus released at a depth of 300 km or more would scavenge through a greater thickness of asthenosphere and would collect more potassium and other associated elements than the water liberated at 150 km.

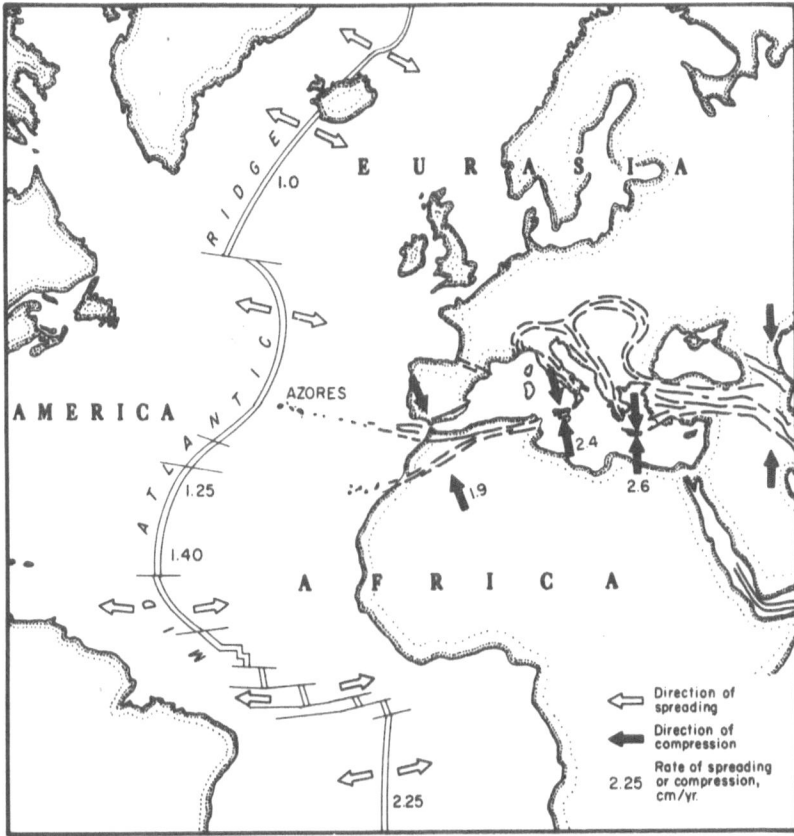

Fig. 17.4. Relative motion between the American, Eurasian, and African plates, resulting in spreading of Mid-Atlantic Ridge and compression and unterthrusting in the Mediterranean Sea. (After Le Pichon 1968)

According to Fyfe and McBirney (1975) phlogopite, having higher P–T stability field than other commonly-occurring hydrous phases, should survive up to a greater depth during the underthrusting of the crust into the mantle and eventually at considerable depths, dehydration of phlogopite should produce potassic magmas. Sodium-rich amphibole, having a lower P–T stability field than phlogopite, should break down at shallower depths and the magmas generated at such depths should thus be relatively richer in sodium than potassium.

Beswick (1976) considered that volcanic activities in the oceanic ridge, continental margin, and crust are produced in different stages of a single whole sequence as shown in Fig. 17.6. He also believed that

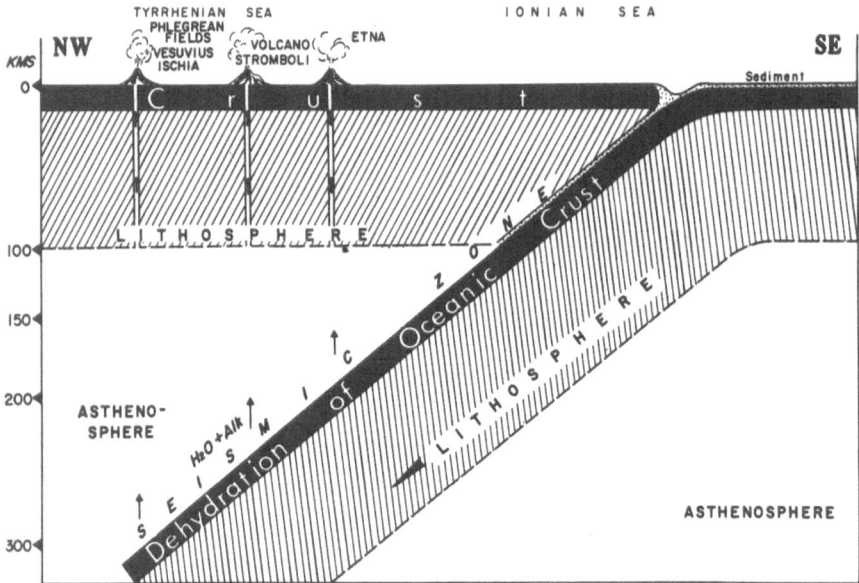

Fig. 17.5. Hypothetical underthrusting of Calabrian arc. (After Ninkovich and Hays 1972)

the main source of K and Rb in the upper mantle is phlogopite. According to his model the subduction of the phlogopite-bearing chondritic mantle may be associated with volcanism in different stages as shown in Fig. 17.6. Beswick concluded that during subduction in such a sequence there should always be a decrease in the production of melt volumes, which should be gradually lower in K content and K/Rb ratio. At each stage, however, K/Rb values of these melts should be appreciably higher than their sources. The model shows that such a process should cause a distinct difference in the K and K/Rb values for suboceanic and subcontinental mantles. He concluded that the suboceanic mantle should be poorer in K and Rb as it descends along the subduction zone at the continental margins and loses its island arc volcanic product to the continental crust and the residual mantle is thus transformed into a highly refractory material with very low K and K/Rb ratio. The new crust thus formed by accretion of volcanic materials in the margin of the continents should consist mainly of island arc materials. Beswick (1976) thus believed that the kimberlites, mica lamprophyres, and ultrapotassic volcanics are derived from a highly refractory and depleted continental mantle material, which lost its

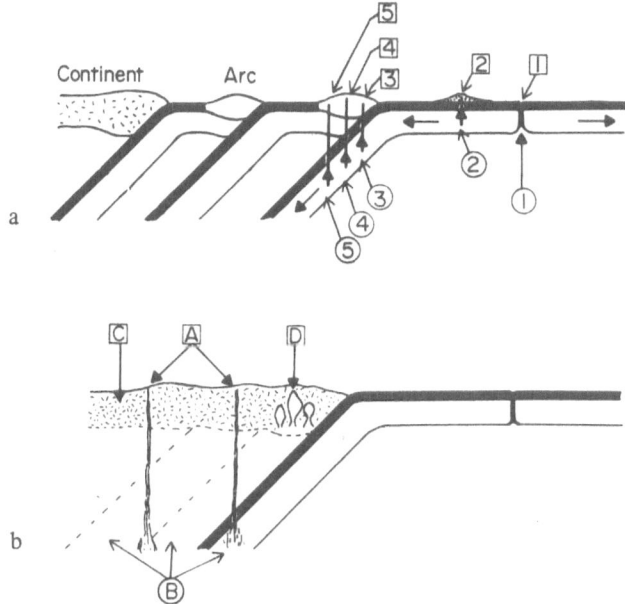

Fig. 17.6 a, b. Schematic representation of *a* the time sequence evolution *(1‑5 in circles)* as successive liquids *(1‑5 in squares)* are lost during transit from ocean ridge to subduction zone; *b* the development of the continental crust *(C)* via island arc accretion with distribution of K and Rb during anatexis of deep crust *(D)*. Late stage addition to the crust of kimberlitic or ultrapotassic material *(A)*, as derived from residual subducted lithosphere. (After Beswick 1976)

andesitic fraction after oceanic or continental margin volcanism. From such a refractory material ultrapotassic magmas may be produced due to partial melting of small amounts of phlogopite-rich fractions, which they have retained. Formation of such a magma should occur not only at the final sequence but should also be characterized by a small volume of melt production.

Wyllie (1973) discussed the maximum stability ranges of hydrous silicate minerals occurring in starting materials and products of a tectonic cycle (Fig. 17.7). He included amphibole, serpentine, talc, muscovite in presence of quartz, phlogopite in presence of olivine, and orthopyroxene in his discussion. The dashed lines in the figure represent the estimated temperatures at the surface boundary of the subducted lithospheric slabs. Wyllie concluded that, "a mantle depleted in a basaltic component beneath an oceanic ridge is not likely to supply serpentine, or talc, or phlogopite to a subducted lithosphere plate. This

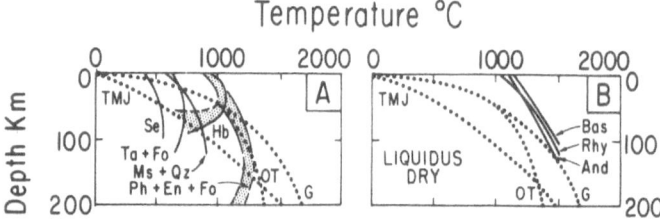

Fig. 17.7. Dehydration and melting reactions compared with estimated temperatures along the upper friction boundary of a subducted lithospheric slab. *Se* serpentine; *Ta* talc; *Fo* forsterite; *Ms* muscovite; *Qz* quartz; *Ph* phlogopite; *En* enstatite; *Hb* amphibole. TMJ, Toksöz et al. (1971). OT, Oxburgh and Turcotte (1970). The diagram is based on experimental results of various workers. (After Wyllie 1973)

suggests that dehydration of lithosphere does not contribute water for magmatic process beyond Dickinson's (1971) arc-trench gap".

17.3.2 Genesis of Magmas of Roman Province

Alverez (1972) discussed the volcanism in the Roman Province in terms of plate tectonics. He considered that 12 million years B.P. the positions of Corsica and Sardinia were different. These microplates then started to rotate, the pole of such rotation was considered by him to be located at 43°22′ North and 9°38′ East. He further thought that the late Miocene and Quaternary volcanism in the region between Pisa and Naples (which includes the Roman Province) was the result of magma, produced along a former Benioff zone, dipping eastward under Italy. The Benioff zone presumably resulted from the subduction of the Corsica-Sardinia plate, which jammed against a trench and ceased rotating. The leading edge of the plate, presumably derived from the thinner crust, broke loose and continued to descend along the subduction zone. A thick sedimentary accumulation occurs towards the east coast of Corsica and Sardinia, which was thought by him to be the fossil trench.

17.3.3 Genesis of Shoshonite

Barberi et al. (1974) considered that the leucite-bearing tephrites from the Vulcanello area of the Eolian arc region were produced from a shoshonitic magma. Some interesting ideas as to the regional structure and depth of formation of shoshonitic magma are therefore discussed below:

Based on the studies of rocks in Fiji (Gill 1970) and New Britain, New Guinea (Jakes and White 1969), the latter workers (Jakes and White 1971) considered that normal alkalic rocks of the island arc region are shoshonites. Miyashiro (1975) believed that the sodium-rich alkalic rocks are more common in the island arc region and therefore suggested that the shoshonitic rocks of Fiji might not have been produced in an island arc environment. Miyashiro thought that the shoshonites of Italy, Fiji, and New Guinea probably are products of hot spots, which may lie somewhat outside the regions of island arcs at depth, or they may be superposed over island arc environments. According to Miyashiro (1975), if the hot spots are of mantle origin (Morgan 1972) then the volcanic products of the hot spots should be independent of shallow structures such as island arcs and continents.

Chapter 18 Generation of Potassium-Rich Mafic and Ultramafic Magma Capable of Producing Leucite-Bearing Rocks

Close geochemical observation of the leucite-bearing mafic and ultra-mafic rocks suggests that any hypothesis to account for their origin must also explain the following features of these rocks:

1. Usually high K_2O content, sometimes as high as 18%.
2. High K_2O/Na_2O ratio, in some cases the K_2O/Al_2O_3 ratio being greater than 1.
3. Lower SiO_2 contents relative to alkali basalts.
4. High enrichment of such trace elements as Rb, Sr, Nb, and Zr (characteristic of salic rocks).
5. High concentration of Ni and Cr (usually associated with ultramafic rocks) in leucitic rocks of some areas.

The hypotheses that have so far been put forward to account for the origin of highly potassic magma capable of producing leucite-bearing rocks can be classified as follows:

1. Hypotheses involving assimilation and contamination.
   a) Shand (1931) and Rittmann (1933): Assimilation of sedimentary carbonate rocks by a granitic or a trachytic magma.
   b) Williams (1936): monchiquite magma + granitic rocks.
   c) Holmes and Harwood (1937) peridotite magma + basaltic rocks.
   d) Gorai (1940): Basaltic magma + granitic rocks.
   e) Holmes (1950, 1965): carbonatite magma + granitic rocks.
2. Hypotheses involving fractionation of magma.
   a) Holmes (1932): separation of eclogite from peridotite; Wade and Prider (1940): generation of mica peridotite magma by separation of alumina-rich rocks from a peridotite melt, followed by differentiation of the mica peridotite.
   b) O'Hara and Yoder (1967): separation of eclogite from a picritic basalt.
3. An hypothesis involving resorption of mica (Bowen 1928).
4. An hypothesis related to gaseous transport (Kennedy 1955).

5. Hypotheses based on partial melting of upper mantle.
   a) Harris (1957): zone melting.
   b) Yagi and Matsumoto (1966) and Gupta et al. (1976): partial melting of phlogopite-bearing ultramafic mantle.

18.1 Hypotheses Involving Assimilation and Contamination

18.1.1 Hypotheses of Shand (1931) and Rittmann (1933)

Daly (1910, 1933) suggested the formation of alkaline magma in general by assimilative reaction between limestone and basaltic magma. While discussing the origin of the syenites and shonkinites from the Palabora area of Transvaal, Shand (1931) supported Daly's hypothesis but considered that their genesis is related to assimilation of limestone by a granitic magma rather than a basaltic magma, followed by expulsion of albite. He considered that further desilication of liquid should cause potassium enrichment and produce such leucite-bearing rock types as orendite, jumillite, venanzite etc. According to Shand, complex reaction procedures observed in the Palabora area should produce a nephelinitic lava at an early stage and a leucitic lava in the later stage.

Rittmann (1933) also believed that the origin of leucite-bearing magma was related to the desilication of a melt rich in alkali feldspar, but he considered that the highly potassic rocks of the Somma-Vesuvius volcanic complex were produced by assimilative reaction between limestone and trachyte. Sinking of ferromagnesian silicates, upward concentration of K_2O by gaseous transfer and differential removal of Na_2O by gases escaping into the wall rock were considered to be additional factors in the evolution of Somma-Vesuvius volcanic complex. According to Rittmann, the hypothesis is further supported by the fact that there is an abundance of xenoliths and fragments of metamorphic limestone in the ejecta of Monte Somma.

Bowen (1922b) objected to the limestone assimilation hypothesis on the grounds that magmas entering into the crust are not superheated, i.e., they do not have large excess of temperature above their crystallization range. He wrote, "A study of some simple equilibrium diagrams, with the object of determining the heat effects connected with solution, gives every reason for believing that the effect is large absorption of heat, usually of the order of magnitude of the latent heat of melting. For simple solution then, it is unquestionable that large amounts of heat will be required" (Bowen 1928; p. 220). He further pointed out that the generation of the feldspathoidal liquid by assimilation process will first require exhaustion of the free silica followed by desilication of the feldspar component. He was of the opinion that the process of assimilation would not be capable of modifying the magma composition to this extent before being used up by crystallization concomitant with solution and desilication.

According to Holmes and Harwood (1937) sedimentary carbonate materials dragged out by the lavas during the eruption of the Bunyaruguru volcano do not show any sign of metamorphic reaction or incorporation. Limestone-bearing formation is known to occur underneath this volcano. In the West Kimberley Province of Australia, limestones are also associated with leucite-bearing rocks, but according to Wade and Prider (1940), there is no field evidence suggesting the resorption of the former. These observations indicate that the parent magmas invading limestone-bearing sedimentary formations in these two localities were not superheated, thus supporting Bowen's (1928) objection.

Wyllie (1974) objected to the limestone assimilation hypothesis on the following grounds.

He pointed out that in synthetic systems there are thermal barriers, which separate silica-rich liquids from liquids capable of precipitating feldspathoids. These thermal barriers do not permit a silica-rich subalkaline melt to cross over to the silica-undersaturated portion even after extensive desilication under isothermal condition. For example, reference to Fig. 18.1 shows that a liquid of initial composition l_1 is in equilibrium with quartz and feldspar of composition f_1. Removal of silica under isothermal condition will cause the bulk composition to move towards M and the liquid and solid composition to l_2 and f_2 respectively. With further desilication, the bulk composition will move finally along the line $l_1 M f_3$ to f_3, where feldspar (f_3) will coexist with only a trace of liquid of composition l_3. With continuous removal of silica, even under disequilibrium condition, the liquid composition should remain within the region $l_1 l_2 l_3$ and will be accompanied by precipitation of feldspars of variable compositions. The presence of a feldspar join (orthoclase-albite, Fig. 18.1) acting as a thermal barrier will prohibit the liquid from going across, thus preventing feldspathoid precipitation unless the temperature of the liquid is higher than the liquidus surface of the albite-orthoclase join (Fig. 18.1). However, such a liquid should not coexist at any stage with crystals.

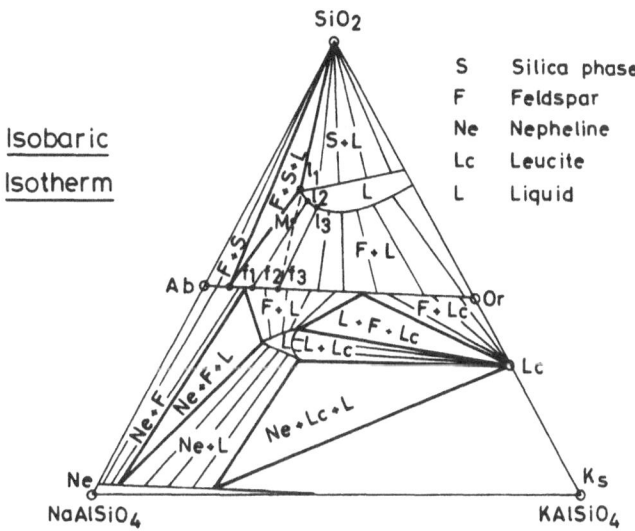

Fig. 18.1. Course of crystallization of liquid in the system $NaAlSiO_4-KalSiO_4-SiO_2$. (After Wyllie 1974)

Desilication of a liquid should always be accompanied by CaO enrichment. Such a situation can be exemplified by considering the system anorthite-nepheline-kalsilite-SiO_2. In this system also the ternary join albite-orthoclase-anorthite acts as a thermal barrier preventing the silica-rich liquid from crossing over to the silica-undersaturated side.

On the basis of available experimental data Wyllie (1974) concluded that there was no indication that the solubility of CO_2 in granitic and feldspathic liquid may lower the liquidus temperatures of the feldspar join sufficiently to destroy the thermal barrier.

Wyllie also pointed out that because CO_2 is only slightly soluble in granitic liquids and very little solid materials are dissolved in a CO_2-rich vapor phase coexisting with a granitic magma, CO_2 may not be an effective transport medium. Study of the system $K_2O-SiO_2-CO_2-H_2O$ (Morey and Fleischer 1940) and $NaAlSi_3O_8-Na_2CO_3-H_2O$ (Koster van Groos and Wyllie 1966, 1968, 1973) suggests that transport of alkalis by vapor or fluid phase may be an effective process if the magma is silica-undersaturated to start with.

Based on the knowledge of trace element contents of trachyte and limestone, Bell and Powell (1969) found that higher concentration of Nb, Sr, and Zr in highly potassic rocks cannot be explained by the assimilation of limestone by a trachyte magma. The assimilation of limestone by a magma low in $^{87}Sr/^{86}Sr$ would also not provide the correlation between Rb/Sr and Sr isotopic ratio.

18.1.2 Assimilation of Peridotite by a Basalt Magma
(Holmes and Harwood 1932)

Holmes and Harwood suggested that the formation of plagioclase-bearing potassic rocks would involve reaction of a peridotite magma with rocks of basaltic composition. They considered the following scheme:

Olivine basalt → trachybasalt → trachyte
Secondary magma (produced by assimilation of a peridotite and a
    basalt magma) → leucite basanite → leucite tephrite
Peridotite or derivative magma → Olivine leucitite → leucitite.

Holmes and Harwood (1937) later found that basaltic rocks are absent in that region; the so-called olivine basalt belonged to a shoshonite-absarokite series. Problems related to the assimilation hypothesis have already been pointed out.

18.1.3 Hypothesis of Williams (1936)

The potassium-rich undersaturated rocks of the Navajo area were considered by Williams to have formed by assimilative reaction between a monchiquite magma and granitic rocks. He provided field evidence of the occurrence of granite-minette association. Williams considered that the primary magma was probably rich in lime, iron, and magnesia, and low in silica, and would thus approximate the composition of monchiquite. Such a magma preferentially dissolved feldspar and ferromagnesian minerals of the granite to produce the potassic rocks of these areas. Williams found that granitic fragments blown out of the Navajo vents show signs of resorption of the feldspars. He thought that the parent monchiquite magma was probably formed by the sinking and selective differentiation of hornblende from a basaltic magma as suggested by Bowen (1928). According to him the monchiquite magma could also be produced from an ultramafic parent by agpaitic type differentiation in which early separation and upward floatation of feldspars and feldspathoids occurred. Presence of fragments of pyrope-bearing rocks, harzburgites and lherzolites in the leucite-bearing rocks of the Navajo area suggested the formation of the primary magma at great depth.

The monchiquite magma which supposedly underwent assimilation was considered by Williams to be a daughter product of either a basalt or a peridotite. It is thus highly unlikely that such a secondary magma could retain enough heat to assimilate granite and then further fractionate to produce various derivative rock types of its own. Absence of granite in Toro-Ankole and Tristan da Cunha also rules out the genesis of leucite-bearing rocks in those areas by this hypothesis.

18.1.4 Hypothesis of Gorai (1940)

Gorai studied the leucite-bearing rocks of Wu-ta-lien-chih of northern Manchuria (Chap. 5.14). In these lavas he noted the presence of granitic xenoliths, the alkali feldspars of which are completely resorbed, whereas quartz and plagioclase grains of these xenoliths are attacked only slightly. On the basis of known chemical compositions of the available rocks in the area, he concluded that leucite-bearing rocks were produced due to selective assimilation by the basalt magma of alkali feldspar and biotite, present in the granite.

Turner and Verhoogen (1960) supported this hypothesis on the basis of the following observations:

1. Presence of granitic xenoliths in many olivine leucitite magma and mica lamprophyres.
2. Close association of mica lamprophyres similar in composition to olivine leucitite with areas of earlier granitic rocks.
3. Association of leucite-bearing rocks with alkali basalts.
4. Availability of such elements as Ba, Rb, Sr, and Zr, also noted in the leucite-bearing rocks.

Yagi and Gupta (1977) made some assimilation experiments between a synthetic granite (alkali feldspar$_{80}$ quartz$_{20}$, Gr) and an alkali diabase (Dia; for analysis see Table 16.2) from Morotu district of Sakhalin. These two rocks were mixed in the following proportions: $Dia_{80}Gr_{20}$, $Dia_{60}Gr_{40}$, $Dia_{40}Gr_{60}$, and $Dia_{20}Gr_{80}$. The mixtures were heated at temperatures between 850° and 1000 °C for 24 to 48 h. In all the samples solidus was reached at 820 ± 20 °C. Above this temperature all mixtures contained sanidine, clinopyroxene, plagioclase, opaque mineral, and glass in variable proportions. In all these runs leucite did not appear, thus suggesting that even if it is assumed that a basaltic magma invading a granitic terrain does have the necessary superheat, the assimilation reaction does not produce any leucite-bearing assemblage.

18.1.5 Carbonatite and Granite Assimilation Hypothesis of Holmes (1950, 1965)

Because the limestone assimilation hypothesis presents various problems, Holmes suggested the formation of the rocks belonging to the O.B.P. series (consisting essentially of olivine, biotite, and pyroxene) and katungite, by reaction between granite and carbonatite magma. These rocks occur in the Birunga volcanic province of Uganda. According to Holmes (1950) carbonatite should not only supply the necessary high temperature, but also gases under high pressure, which were probably responsible for drilling vents, production of tuffs and lapilli, and charging emanations. The older volcanic tuffs were found by Holmes to be rich in the cementing materials. Holmes (1950) found the lava flows of the Fort Portal area to be highly carbonated. Combe and Holmes (1945) also described the presence of travertine material in the Katwe and other associated craters. Occurrence of carbonatites from east Africa has been described by Tuttle and Gittins (1966), Heinrich (1966), and Le Bas (1977). Holmes (1950) described his hypothesis by the following simplified equation:

$$C + G = O.B.P. + E, \tag{18.1}$$

where C stands for carbonatite, G for granite, and E for volcanic emanations. A portion of O.B.P. may react with E to produce katungite (K):

$$K = A \cdot (O.B.P.) + E . \tag{18.2}$$

In Eq. (18.2), A is a constant. If Eq. (18.2) is subtracted from Eq. (18.1), then

$$C + G - K = O.B.P. - A (O.B.P.)$$

or

$$C + G = K + O.B.P. (1-A) . \tag{18.3}$$

In terms of weight percentage Eq. (18.3) can be modified as follows:

$$M.K. + (100-M) O.B.P. = N.C. + (100-N) \cdot G . \tag{18.4}$$

In Eq. (18.4), M and N are constants. By knowing the average composition of the granite, the katungite and the rocks of the O.B.P. series

Holmes (1950) determined the values of M and N. If these values are substituted in Eq. (18.4), Eq. (18.5) can be written as follows:

$$60.9 \text{ K} + 39.1 \text{ O.B.P.} = 51.9 \text{ G} + 48.1 \text{ C} . \tag{18.5}$$

For the genesis of proto katungite (P.K.) values of the constants are different and the equation of their formation is as follows:

$$39.5 \text{ P.K.} + 60.5 \text{ O.B.P.} = 53.2 \text{ G} + 46.8 \text{ C} . \tag{18.6}$$

Using these equations Holmes (1950) determined the composition of the carbonatite magma. He found field evidence of marginal alteration of biotite relics in katungites. In rocks of the ugandite and mafurite series, reaction rims are quite common. Analyses of such rims in kalsilite ugandite showed that they corresponded closely to a biotite pyroxenite slightly defficient in Al_2O_3, MgO, K_2O, and H_2O.

The hypothesis of Holmes (1950) was supported by Higazy (1954), who noted that carbonatites are rich in such elements as Sr, La, Y, and Ba. Granites on the other hand are rich in Rb, Ba, Zr, and Ga. Assimilation of granite and carbonatite magma therefore should produce a melt rich in all those elements. However, like granite, carbonatites are also found to be poor in Ni and Cr. The high concentration of these two elements in potassic rocks cannot be explained by this hypothesis alone. Higazy suggested that the original carbonatite magma was probably magnesia-rich and contained a higher amount of Ni and Cr, but after assimilative reaction the product became enriched in Ni and Cr, whereas the residual carbonatite became impoverished in these two elements. Higazy considered that ultramafic rock types of the area such as katungites were produced by reaction of a considerable amount of magnesium-rich carbonatite with a small amount of granitic material. Formation of rocks such as glimmerite probably would need reverse quantities of the above two parent materials. In the genesis of melilite-nepheline leucitites, leucite kivites and nepheline melilitites, which are relatively low in Ni and Cr, a calcium-rich carbonatite magma was probably involved. Thus by varying the proportion of granite and carbonatite composition, various types of potassium-rich volcanic rocks can be produced.

On the basis of their Sr isotopic study of the carbonatites from East Africa and leucite-bearing rocks of central and equatorial Africa, Bell and Powell (1969) also found this hypothesis acceptable.

Koster van Groos and Wyllie (1966) showed that in the system $Na_2O-Al_2O_3-SiO_2-CO_2$ carbonate and sodium-aluminum silicate liquids are immiscible over a certain temperature range. Thus the formation of carbonatite from another primitive magma by liquid immiscibility is possible. The lava flow of Oldoinyo Lengai of Tanganyika (Dawson 1966) provides such an example. Roedder (1965) found the presence of glass (presumably of basaltic composition) in the materials of olivine-bearing nodules from 66 localities throughout the world.

It should be pointed out here that except for few instances such as Kaiserstuhl (Wimmenauer 1974), West Eifel (Lloyd and Bailey 1969) and the Tundulu region of Malawi (Gittins 1966) carbonatites are generally found in association with fenites and sodium-rich undersaturated rocks, such as ijolites, nephelinites etc., rather than leucite-bearing volcanic rocks. In most other localities leucitic rocks occur independently of carbonatites.

In contrast to limestone, carbonatites are usually richer in such elements as Ba, Sr, La, and Y, otherwise from a physicochemical stand point there is not much difference between the limestone assimilation hypothesis of Shand (1931; Chap. 18.1.1) and the carbonatite assimilation hypothesis of Holmes (1950, 1965). In the former, the source of heat was considered to be a granitic magma, whereas in the latter the supplier of heat was considered to be a carbonatite magma. Thus the objection raised by Bowen (1928) and Wyllie (1974) against the limestone assimilation hypothesis should also apply to the carbonatite assimilation hypothesis. The mechanism suggested by Holmes (1950) should be effective only if the parent carbonatite magma is superheated sufficiently to desilicate free silica and feldspar, while remaining at a temperature above the liquidus surface of the albite-orthoclase thermal barrier (for detailed arguments see Chap. 18.1.1). This is highly unlikely especially in the light of the fact that carbonatites present such genetic features which suggest that they themselves are of derivative origin. The features suggesting late stage formation of carbonatite have been summarized by Heinrich (1966) as follows:

a) The volumes of even the largest carbonatites are negligible compared to those of alkalic rocks.

b) The mineralogy and the stability fields of their synthetic equivalents suggest a low temperature of formation. Dawson (1966) studied the lava flow of the active carbonatite volcano of Oldoinyo Lengai of Tanzania and concluded that it was unlikely in view of the unusual chemistry of the lava that the temperature was greater than 500 °C.

c) The presence of such a sequence as nepheline syenite → silico-carbonatite → carbonatite containing accessory silicates.

d) The presence of rocks such as sövites and rauhaugite, grading to complex carbonatites, which contain assemblages with hydrothermal characteristics, such as fluorite, barite, and sulphide.
e) Many alkalic rock complexes occur without carbonatite association, but only very rarely are carbonatites found without alkalic rocks.

18.2 Hypotheses Involving Fractionation

18.2.1 Views of Holmes (1932) and Wade and Prider (1940)

Holmes suggested that leucite-bearing volcanic rocks were produced by differentiation of a primary magma under higher pressures. He suggested that separation of eclogite and dunite from a primary peridotite magma under high volatile pressure should produce kimberlites; if, however, the volatiles escape, the resulting rocks would be similar to olivine leucitite. Eclogites are usually high in the Na_2O/K_2O ratio and relatively rich in silica. Peridotites are generally richer in sodium content than potassium. Extraction of eclogite may therefore produce a potassium-rich residual liquid. Holmes (1932) instanced the presence of eclogite in kimberlites in support of his views. High-pressure differentiation was favored by him for the following reasons:

a) Presence of tuffs and explosion craters, which would require high concentration of volatiles in the magma.
b) The uplift in the Ruwenzori area would need intense lateral compression at a much greater depth. Gravity anomaly data of Bullard (1936) are in agreement with this uplift.

Later work of Holmes and Paneth (1936), however, showed that the eclogite xenoliths are not cognate. Eclogites are of Precambrian age, whereas the emplacement of kimberlite took place in the Cretaceous. Therefore the eclogite and the kimberlite are not related petrogenetically.

Wade and Prider (1940) considered that mica peridotite magma was responsible for the genesis of leucite-bearing rocks of the West Kimberley province of Australia. The formation of such a mica peridotite melt was thought by them to be due to the separation of kyanite or garnet-bearing alumina-rich rocks from a peridotite magma. Further separation of olivine from the mica peridotite would produce a residual liquid, similar in composition to orendite with all minor constituents, and relatively rich in K_2O and poor in Na_2O. They further postulated that separation of phlogopite and diopside from the mica peridotite liquid would produce a melt poor in iron oxide, lime, and magnesia. Partial resorption of phlogopite by such a magma would produce a highly potassic residual liquid, which by now should be slightly richer in alumina and silica. According to Wade and Prider, wolgidites were produced from such a magma at considerable depth. Madupite was considered by them to be a direct derivative of an orendite magma, whereas wyomingite was considered to be a daughter product of a madupite magma.

If it is considered that peridotite constitutes the upper mantle material, generation of a peridotite magma would require complete melting of such material, involving large enthalpy of melting. Most earth scientists now believe that the generation of magma in the upper mantle is related to partial melting rather than complete melting of the upper mantle materials. In the generation of a mica peridotite magma from peridotite, subtraction of alumina-rich rocks from a peridotite melt was suggested by Wade and Prider (1940). In the leucite-bearing rocks of West Kimberley such xenoliths are also not observed. It is interesting to note that Holmes and Harwood (1937) and Wade and Prider (1940) realized the importance of mica-bearing ultramafic rocks in relation to the genesis of highly potassic undersaturated rocks. Stability of phlogopite under high P–T conditions, equivalent to that of the upper mantle, was not known to these petrologists, thus, although they realized the close genetic connections between mica-bearing ultramafic rocks with leucite-bearing volcanics, they speculated the origin of the former at shallower depth than the mantle. Such a mica peridotite magma would not account for the observed radiogenic Sr as noted in the leucite-bearing rocks of central and equatorial Africa.

18.2.2 Eclogite Fractionation Hypothesis of O'Hara and Yoder (1967)

High pressure fractionation of eclogite from a primary picrite magma at depth (equivalent to 25 and 40 kb) was considered by O'Hara and Yoder to be responsible for the formation of a liquid with a high K_2O/Na_2O ratio. Such a liquid should ultimately produce kimberlites or related highly potassic rock types. On the basis of their Principal Latent Vector diagram of leucite-bearing rocks from the Bufumbira region,

Fig. 18.2. Principal Latent Vector Variation diagram showing possible compositions of source rocks and derivative liquids in relation to basanites; phonolitic tephrites, and leucites are numbered *1* to *3* respectively; U_{1-4} represent liquid compositions of phonolitic tephrites, leucites, and basanites. *P* field of pyroxenite and biotite pyroxenite compositions. *L* field of spinel and garnet lherzolite compositions; *f* pyrolite III composition (Green and Ringwood 1963); E_{1-3} Gleneg, Loch Duich, and Salt Lake eclogite compositions respectively (Yoder and Tilley 1962); *W-X-Y-Z* garnet-clinopyroxene-orthorhombic pyroxene-olivine compositions from garnet peridotite (eclogite; O'Hara and Yoder 1967); *O-R-Q* phlogopite-chrome diopside-olivine plane (at 1 wt.% chromite), i.e., the chemical mode of average phlogopite compositions *(MP)*; *PL* possible primary liquid composition. (After Ferguson and Cundari 1975)

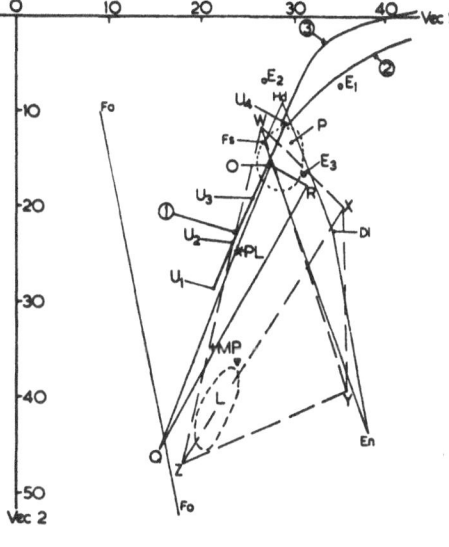

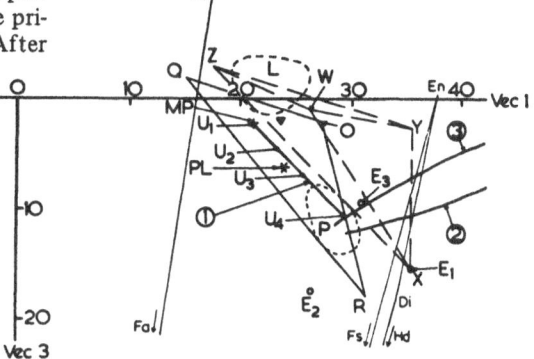

Ferguson and Cundari (1975) concluded that fractionation of eclogite from a picritic liquid may not produce the reported trends of Bufumbira lavas. Figure 18.2 is taken from Ferguson and Cundari (1975), which indicates that E_1 to E_3 represent eclogitic compositions. Subtraction of these compositions or an assemblage consisting of garnet, clinopyroxene, and olivine (W, X, Y, and Z respectively, Fig. 18.2) would cause the liquid to move to the left (Fig. 18.2), thus producing a liquid of more primitive composition. Separation of garnet peridotite from U_3 or P1 should move the composition of the melt to the right, producing a more evolved composition; the observed trend of crystal-

lization thus does not correspond to that noted in the case of Bufumbira basanites. According to Ferguson and Cundari such a process does
not produce enrichment of potassium in the residual liquid.

A nepheline-normative and a hypersthene-normative picrite were
studied under various P_{H_2O}-T conditions by Gupta and Yagi (1977).
Their study (Chap. 16.4) showed that separation of an eclogite fraction from a nepheline-normative picrite at 25 kb P_{H_2O} and 800 °C
does not produce a liquid with a K_2O/Na_2O ratio higher than 1. The
liquid formed this way is in effect very similar to a nephelinite. They
also found that after crystallization of garnet and pyroxene from a
hypersthene-normative picrite at 25 kb P_{H_2O} and 1000 °C, the residual
liquid is equivalent to some basalts (low in K_2O, and K_2O/Na_2O ratio
< 1). The hypothesis of O'Hara and Yoder (1967) thus does not provide
a suitable mechanism for the genesis of leucite-bearing mafic and ultramafic magma.

18.3 Fractional Resorption of Mica and Amphibole (Bowen 1928)

Bowen considered that after crystallization of olivine, pyroxene, and
some plagioclase from a basaltic liquid, hornblende and/or biotite may
start to crystallize in the residual liquid and later the amphibole and/or
mica may sink into the hotter zone of the magma. He thought that the
reaction of amphibole with the liquid should produce a sodium-rich
undersaturated magma, whereas the resorption of mica should produce
a potassium-rich liquid. A process of "squeezing out" was further
envisaged by him to maintain the strongly undersaturated character of
the alkalic liquid.

While studying the system $KAlSiO_4$-Mg_2SiO_4-SiO_2-H_2O, Luth
(1967) noted that at pressures below 500 bars phlogopite does not coexist with water-saturated liquid, whereas at pressures greater than
1200 bars it is in equilibrium with a variety of water-saturated liquids.
Luth (1974) used polybaric isothermal sections (Fig. 18.3a-c) to il-
lustrate phlogopite resorption. In these sections crystallization paths
of the liquids of bulk compositions M (Fig. 18.3a), N (Fig. 18.3b) and
A (Fig. 18.3c) are shown by dashed lines, whereas the double-dashed
lines represent the portion of the system where resorption of phlogopite takes place. In the case of the 900 °C isothermal section a major
part of the resorption is associated with the disappearance of the li-

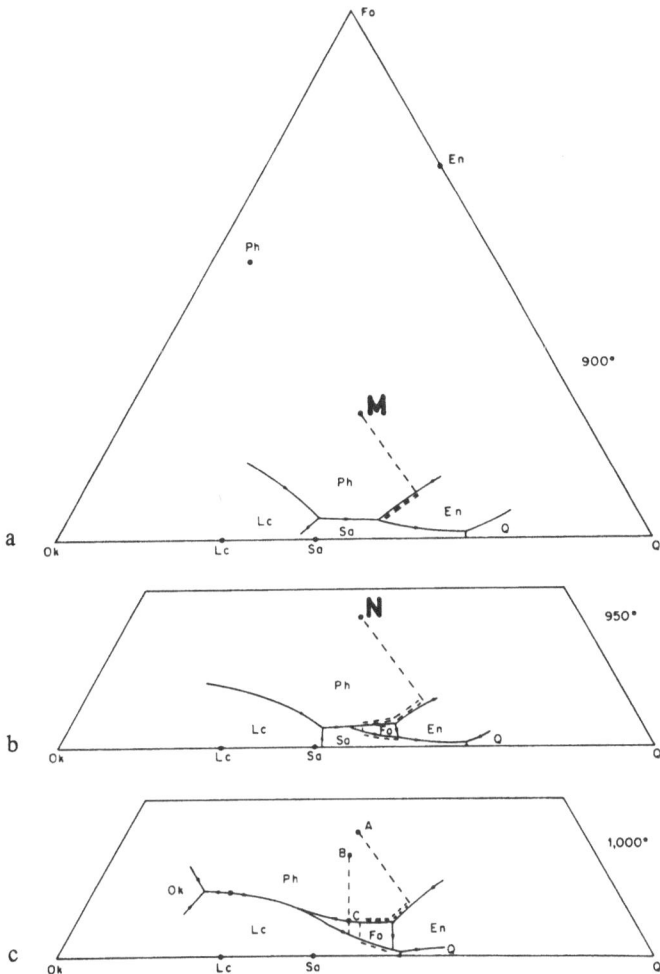

Fig. 18.3 a–c. Isothermal polybaric projections of the vapor-saturated liquis in the system $KAlSiO_4-Mg_2SiO_4-SiO_2-H_2O$. *Dashed and double-dashed curves* refer to liquid compositions discussed in the text. Three figures refer to isothermal sections at *a* 900 °C, *b* 950 °C, *c* 1000 °C. (After Luth 1974). See also List of Abbreviations

quid, whereas at higher temperatures, reaction of mica takes place in the presence of considerable amounts of liquid. The illustrations of phlogopite resorption in Fig. 18.3 a, b represent equilibrium crystallization, whereas Fig. 18.3 c may be used to study resorption of phlogopite associated with disequilibrium crystallization. If enstatite, produced by isothermal resorption of phlogopite between 3400 and 2500 bars

(Fig. 18.3c) is separated from the liquid, the bulk composition should move from A to B, which should contain 71% liquid and 29% phlogopite. As the pressure drops phlogopite is further resorbed and forsterite starts to crystallize. If this forsterite is again removed from the liquid the bulk composition should move to C, where complete disappearance of phlogopite takes place. With further drop in pressure forsterite continues to crystallize.

Luth (1974, p. 509) considered that his experimental results supported the views of Bowen (1928). It should be pionted out, however, that in his illustrations of the resorption of mica by the residual liquid, the initial bulk compositions M, N, and A (Fig. 18.3a–c) lay in the system $KAlSiO_4$–Mg_2SiO_4–SiO_2–H_2O and were thus already enriched in K_2O. The composition of liquid with which phlogopite reacts is also quite different from that suggested by Bowen (1928). The main objective of Bowen's hypothesis was to generate a K_2O-rich liquid from a basaltic magma, which is relatively poor in K_2O. A process of filter-pressing is also essential in Bowen's mechanism to maintain the strongly undersaturated character of the melt. Luth's illustrations merely describe the fact that phlogopite appears at a higher P_{H_2O} (> 2 kb) from a potassium-rich ultramafic liquid but at low pressures mica reacts with liquid to produce magnesium orthosilicates and feldspathoids. Minor variations of the K_2O and SiO_2 content of the liquid are also shown in his illustrations.

Bell and Powell (1969) pointed out a weakness of the Bowen hypothesis on the basis of their Sr isotopic studies. They showed that for the process to explain the formation of leucite-bearing rocks, the mica would need to remain in the liquid for a long period, possibly a million years, for the observed $^{87}Sr/^{86}Sr$ ratio of 0.711 to develop. They also pointed out that assimilation of biotite by basalt should not produce the high concentrations of Nb, Zr, and Sr, and the lack of correlation between the ^{87}Sr, $^{87}Sr/^{86}Sr$ ratio and K content that characterize the leucite-bearing rocks.

18.4 Zone-Refining Hypothesis of Harris (1957)

Harris thought that a considerable portion of the subcrustal mantle material is close to the melting temperature. A body of magma, produced on a small scale locally at some depth, would move upward by simultaneous melting and mixing of the rocks in the roof with crystallization and deposition at the bottom of the magma chamber. In such a zone-melting process, elements which are more soluble in the liquid than the crystalline materials would be retained in the liquid zone. Harris considered that by such a process minor elements such as K, Ba, Rb, Zr, and others, which are unable to substitute in the mineral phases such as olivine, pyroxene, and spinel, would be concentrated in the liquid. The removal of such elements would depend on the distribution through which the molten zone moves and also the distribution coefficient K of the element between the solid and liquid phases of the material. Thus,

$$\text{K} = \frac{\text{Concentration of element } A \text{ in the solid phase}}{\text{Concentration of element } A \text{ in the liquid phase}}.$$

If the solid material contains t ppm of an element A and if $\text{K} = \frac{1}{30}$, then the liquid produced by melting would contain $\frac{1}{30}$ xt ppm of A.

Thus Harris (1957) considered that continuous melting of the original solid material and crystallization of the purified phases behind would result in the concentration of A in the liquid. The process is thus an ingenious method to produce potassium-rich mafic and ultramafic magma rich in K and certain other minor elements. Turner and Verhoogen (1960) argued that if the process is considered to be the normal mode of formation of magma, then potassium-rich mafic and ultramafic rocks should be the rule rather than exception. Bell and Powell (1969) considered that zone melting is a continuous rather than a spasmodic process, and thus such a mechanism does not explain the variation in the concentration of isotopic Sr observed in the leucite-bearing rocks of central and equatorial Africa. Contamination of the magma with crustal rocks is thus necessary to explain this variation in Sr isotopic composition. The zone-refining hypothesis is thus not independent of assimilation and, therefore, may not be considered as an effective mechanism.

18.5 Genesis of Potassic Rocks by Volatile Transport

Lindgren (1933) considered that differentiation largely through gaseous transport produced trachytes, phonolites, basanites, and many other volcanic rocks. Rittmann (1933) also suggested this mechanism together with assimilation of limestone by trachyte (Chap. 18.1.1) to be an effective way of producing the leucite-bearing rocks of Vesuvius. Morey and Hesselgesser (1952) showed that 40% by weight of alkali silicates was dissolved by the aqueous vapor phase at 400 °C and 2200 bar. From these observations Kennedy (1955) considered that the high solubility of alkali silicates in gaseous water indicates that they may strongly associate with water molecules and their solubility is very much dependent on pressure. Thus when there is release of pressure, resulting from fracturing during the escape of volatiles, alkalis should be left behind because of the decrease of solubility. Concentration of alkalis at the top of an igneous stock may result from such a process, as in the case of potash enrichments in igneous rocks in Jumbo basin, Alaska (Kennedy 1953) or the formation of pseudoleucite crystals in the contact zone of a shoshonitic dike (Larsen and Buie 1941).

These processes suggest the formation of highly alkalic rocks in the final stages of igneous activity. However, by such a process alkalic rocks should be formed in an isolated cupola with surface connections during the dying stage of magmatism. It is also difficult to see how potassium-rich lavas can emerge through vents after eruption of normal molten rocks by such a process. The process may be considered to account for the enrichment of alkalis in the roof and wall rocks of a magma chamber or near the vicinity of an intrusive body, but concentration of alkalis by such a process should be observed only locally, as found by Larsen and Buie (1941). This mechanism alone may not be held responsible for the generation of large volumes of potassium-rich mafic and ultramafic lavas, as it would require the existence of a large source of batholithic dimension, always associated with leucite-bearing lavas, to supply the large quantities of alkalis.

18.6 Generation of Potassium-Rich Mafic and Ultramafic Magma by Partial Melting of Phlogopite-Bearing Ultramafic Upper Mantle

Experimental results presented in Chap. 15 show that a melilite-nepheline leucitite, a leucite basanite, and a katungite were transformed at high pressures (equivalent to those of upper mantle) into a richterite-phlogopite pyroxenite, a phlogopite-garnet pyroxenite and a phlogopite pyroxenite, respectively. A natural tephrite and a basanite (both containing leucite and nepheline) convert at 750 °C and 5 kb P_{H_2O} to garnet-richterite-phlogopite pyroxenite (± nepheline ± magnetite). It is interesting to note here that in all high pressure assemblages (> 10 kb) the dominant phases are phlogopite and pyroxene with small amounts of richterite or garnet. Even though olivine was present in the starting materials of the two basanites and the katungite, it disappears at high pressures. In case of the two basanites, absence of olivine at high pressures is related to its reaction with leucite plus liquid and vapor to form phlogopite, whereas in the case of the katungite, olivine reacts with akermanite and liquid to produce diopside and phlogopite [Eq. (15.1), Chap. 15.2].

In the experiments on leucite-bearing rocks, if olivine were present in all. the starting materials in larger proportions than the amount of phase with which it has a reaction relationship, olivine would have survived at high pressures and the resulting assemblages might have corresponded to a phlogopite peridotite containing richterite or garnet. For example, in their study of a kalsilite mafurite glass with considerable amounts of normative olivine, Edgar et al. (1976) noted its conversion to phlogopite peridotite at pressures equivalent to that of the upper mantle.

Phlogopite-bearing peridotite and pyroxenite xenoliths found within leucite-bearing lavas of various localities (Chaps. 5.1 and 14.3) may thus represent the parent materials for potassium-rich mafic and ultramafic magma capable of producing leucite-bearing rocks.

Evidence from extraterrestrial sources and geophysical and petrological studies suggest that although the upper part of the mantle is inhomogeneous, it is in general composed of peridotitic material (for detailed discussion see Wyllie 1971, pp. 114-120; Yoder 1976; pp. 12-43). Using variable sources Wyllie (1970) constructed a generalized phase diagram of an estimated mantle peridotite, which suggests that at a depth of about 55 km (below the ocean floor) to 70 km (below the continents), the mantle mineralogy is similar to that of a garnet peridotite; a mantle of such mineralogical composition has also been suggested by Green (1964), Carswell (1968), Kushiro and Yoder (1969), and Anderson (1970).

The presence of phlogopite as an accessory phase in the peridotitic upper mantle has already been suggested by Yagi and Matsumoto (1966), Yoder and Kushiro (1969), Modreski and Boettcher (1972, 1973), Forbes and Flower (1974), and Gupta et al. (1976). It is therefore highly possible that partial melting of such a phlogopite-rich portion of the garnet peridotite (± richterite) mantle will produce a potassium-rich mafic and ultramafic liquid. Discussion in Chap. 15.2.4 suggests that the amount of garnet in the phlogopite peridotite should determine the plagioclase content, whereas the concentration of richterite should control the Na_2O content of the magma. However, if the potassic liquid is generated at a depth greater than that of the upper stability limit of richterite (<100 km, see Chap. 14.4.1), its sodium content should be controlled by the amount of jadeite molecule present in the clinopyroxene of the phlogopite-garnet peridotite (Chap. 15.2.4).

It has been pointed out at the beginning of this chapter that a successful hypothesis dealing with the genesis of highly potassic magma should explain why leucite-bearing rocks are richer not only in Rb, Sr, Ba, Li, and Zr (characteristic of salic rocks), but also in Ni and Cr (characteristic of ultramafic rocks). It should now be examined from the geochemical aspect whether the partial melting of a phlogopite-bearing ultramafic mantle could produce a liquid rich in these elements.

The elements such as Rb, Sr, Ba, and Zr are considered to be readily acceptable in the structure of mica (Ahrens 1965; p. 351). Ni is usually associated with olivine, where it is present as the Ni_2SiO_4 molecule (Carmichael 1967), whereas Cr is contained in garnet (Sobolev et al. 1975a) as knorringite ($Ca_3Cr_2Si_3O_{12}$, Ringwood and Kesson 1977). Cr may also enter into the crystal structure of clinopyroxene at high pressure (Irving 1974). At atmospheric pressure diopside ($CaMgSi_2O_6$) incorporates considerable amounts of kosmochlore molecule ($NaCrSi_2O_6$) in solid solution (Ikeda and Yagi 1972). Thus if the partial melting of phlogopite-garnet peridotite produces a crystal-liquid mush, the liquid should be relatively enriched in Rb, Sr, Ba, and Zr, whereas the relict crystals should have higher concentration of Ni (in olivine) and Cr (in garnet and clinopyroxene). The bulk chemistry of the crystal-liquid aggregate will thus be characterized by enrichment of elements, typical of not only salic rocks but also of ultramafic rocks.

Gast (1968) and Griffin and Murthy (1969) showed that the Rb/Sr ratio of the liquid produced by partial melting of garnet peridotite containing phlogopite may also be similar to that of the potassic lavas of the Roman Province. Formation of a leucite-bearing mafic and ultramafic liquid by partial melting of potassium-rich substratum is also consistent with the results of the Sr isotopic study of Bell and Powell (1969).

The variation in $^{87}Sr/^{86}Sr$ and Rb/Sr ratios in leucite-bearing rocks has often been considered as a proof that potassic magmas have been produced by different degrees of crustal contamination. However, as suggested by Faure and Powell (1975), this variation may be explained by one of the following reasons:

a) The average Rb/Sr ratio of the upper mantle may decrease at greater depth thus the $^{87}Sr/^{86}Sr$ ratio will be dependent on the depth of formation of the magma.

b) Individual mantle phases may remain isotopically closed for a long period of time, thus developing different $^{87}Sr/^{86}Sr$ ratios.

c) The Rb/Sr ratio of the upper mantle may vary laterally and therefore magmas formed at different subcrustal sites over a long period of time should have different Rb/Sr ratios.

d) Because of variable Rb/Sr ratios in different portions of a long-lived magma chamber, melts of different $^{87}Sr/^{86}Sr$ ratios may be produced (Artemov and Yaroshevskiy 1965).

Thus the idea of formation of a highly potassic magma by partial melting of a phlogopite peridotite-bearing upper mantle is supported by experimental, field, and geochemical observations.

18.7 Summary and Conclusions

A review shows that different hypotheses for the generation of leucite-bearing magmas by assimilation between various rocks and magma pairs have been very popular in the past. In all these hypotheses formation of leucite has been related to the desilication of alkali feldspar present in either a granite or a trachyte magma or in the granitic country rocks.

It is, however, noted that assimilation of foreign materials by a magma should be associated with large absorption of heat. Field evidence suggests that magmas invading sedimentary formation or salic country rocks do not have sufficient excess of temperature above their crystallization range. Thus assimilation processes may not be capable of modifying magma composition before being used up concomitant with solution and desilication.

Experimental studies of silicate systems show the presence of thermal barriers separating silica-rich liquids, from liquids capable of producing feldspathoids. These prohibit the formation of leucite-bearing magma by desilication of feldspar due to assimilative reaction between limestone and trachytic or granitic magma involving crystal-liquid reaction. Assimilation processes involving limestone and carbonate magma also would lead to release of CO_2 contributing to severe loss of heat and subsequent crystallization of the magma.

Observed field relations do not support the idea of formation of leucite-bearing magma by separation of eclogitic materials from a peridotitic magma. Study of variation diagrams and experimental evidence indicates that subtraction of eclogite from a picritic magma also does not produce a magma rich in K_2O with a K_2O/Na_2O ratio greater than unity.

The Sr isotopic studies of leucite-bearing rocks do not support the idea of formation of highly potassic magmas by either zone refining hypothesis or by reaction between mica and basaltic liquid.

Generation of leucite-bearing rocks by gaseous transfer mechanism alone does not seem to be too attractive. This process may be considered to account for the enrichment of alkalis in the roof of a magma chamber or near the vicinity of an intrusive body, but for the formation of large volumes of leucite-bearing magmas, the process would require a large source of batholithic dimension always associated with leucitic rocks to supply large quantities of alkalis.

There is, however, enough field and experimental evidence to suggest that parent materials for leucite-bearing mafic and ultramafic magma may be represented at depth by phlogopite-garnet peridotite (± richterite). The presence of such an upper mantle material is consistent with both geophysical and petrological findings.

Trace element geochemistry of phlogopite-garnet peridotite suggests that partial melting of such mantle materials may produce a crystal-liquid aggregate, which should be enriched in such trace elements as observed in leucite-bearing rocks.

Observed $^{87}Sr/^{86}Sr$ ratios of leucite-bearing lavas can be explained if it is assumed that they are produced by partial melting of an old phlogopite-bearing upper mantle.

The chemistry and mineralogy of the phlogopite-bearing ultramafic upper mantle may not be uniform both vertically or laterally, thus having different Rb/Sr ratios. Therefore if highly potassic magmas are produced at different depths or at different sites in the same level over a long period of time, lavas of different $^{87}Sr/^{86}Sr$ ratios will be produced.

It is therefore concluded that the generation of potassic magma is related to partial melting of phlogopite-rich garnet peridotite (± richterite). Such liquids may still hold relict crystals of phlogopite, garnet, olivine, pyroxene, and richterite and ascend toward the surface and convert to leucite and melilite (or plagioclase)-bearing assemblages (± nepheline) near the volcanic or subvolcanic regime. Formation of further varieties of leucite-bearing rocks may be related to fractional crystallization of the magma at shallow depth prior to their eruption.

References

Abbott MJ (1969) Petrology of the Nandewar Volcano, N.S.W. Contrib Mineral Petrol 22:115-134

Ahrens LH (1965) Distribution of the elements in our planet. McGraw Hill Inc, New York, p 110

Ahrens W (1961) Die Tektonische Stellung des Laacher See-Vulkanismus. Fortschr Mineral 39:93-95

Alverez E (1972) Rotation of the Corsica-Sardinia microplate. Nature (London) Phys Sci 235:103-105

Ambrosetti P, Azzaroli A, Banadonna FP, Follieri M (1972) A scheme of Pleistocene chronology for the Tyrrhenian side of Central Italy. Boll Soc Geol Ital 91:169-184

Anderson DL (1970) Petrology of the mantle. Mineral Soc Am Spec Pap 3:85-93

Anderson O (1915) The system anorthite-forsterite-silica. Am J Sci 4th Ser 39:407-454

Aoki K (1974) Phlogopite and potassic richterites from mica nodules in south African kimberlites. Contrib Mineral Petrol 48:1-7

Appleton JP (1972) Petrogenesis of potassium-rich lavas from the Roccamonfina volcano, Roman region, Italy. J Petrol 13:425-456

Artemov YuM, Yaroshevskiy AA (1965) The isotopic composition of strontium as an indicator of character and duration of magmatic differentiation. Geochim Cosmochim Acta 2:810-813

Bailey DK (1964) Crustal warping - a possible tectonic control of alkali magmatism. J Geophys Res 69:1103-1111

Bailey DK (1974) Continental rifting and alkaline magmatism. In: Sørensen H (ed) The alkaline rocks. John Wiley and Sons, London New York Sydney Toronto, pp 148-159

Baker PE, Rea WJ (1978) Compositional variations in the plateau basalts. Int Geodynamics Conf Western Pacific and Magma Genesis, Tokyo. Extended abstract 206-207

Baker PE, Gass IG, Harris PG, Le Maitre RW (1964) The volcanological report of the Royal Society expedition to Tristan da Cunha. Philos Trans R Soc London Ser A 256:439-578

Banerjee S (1953) Petrology of the lamprophyres and associated rocks of Raniganj coal field. Indian Mineral J 1:9-29

Bannister PA, Hey MH (1931) A chemical, optical, and X-ray study of nepheline and kaliophyllite. Mineral Mag 22:569-578

Bannister PA, Hey MH (1942) Kalsilite, a polymorph of $KAlSiO_4$ from Uganda. Mineral Mag 26:218-224

Bannister PA, Sahama ThG, Wiik HB (1953) Kalsilite and venanzite from San Venanzo, Umbria, Italy. Mineral Mag 30:46-58

Barberi F, Innocenti F, Ferrara G, Keller J, Villari L (1974) Evolution of Eolian arc volcanism (southern Tyrrhenian Sea). Earth Planet Sci Lett 21:269-276

Barbieri M, Penata A, Turi B (1975) Oxygen and strontium isotope ratio in some ejecta from the Alban Hills volcanic areas, Roman Comagmatic region. Contrib Mineral Petrol 51: 1327-1333

Barth TFW (1932) The structure of the mineral of the sodalite family. Z Krist 83: 405-410

Barth TFW (1933) Pyroxene von Hiva Oa, Marquesas-Inseln und die Formel titanhaltiger Augite. Neues Jahrb Mineral Beil 64: 217-224

Barth TFW (1963) The composition of nepheline. Schweiz Mineral Petrogr Mitt 43: 153-164

Bell K, Powell JL (1969) Strontium isotopic studies of alkalic rocks: The potassium-rich lavas of the Birunga and Toro-Ankole regions, east and central equatorial Africa. J Petrol 10: 536-572

Beswick AE (1976) K and Rb relations in basalts and other mantle-derived minerals. Is phlogopite the key? Geochim Cosmochim Acta 40: 1167-1193

Bird GM, Anderson GM (1973) The free energy of formation of magnesian cordierite and phlogopite. Am J Sci 273: 84-91

Black R, Girod M (1968) Controle structural du volcanisme ancient et recent deins less regions du Hogger, Air Nigeria et Cameroun. Proc Geol Soc London 1644: 263-266

Black R, Girod M (1970) Late Palaeozoic to recent activity in West Africa and its relationship to basement structure. In: Clifford TN, Gass IG (eds) African magmatism and tectonics. Oliver and Boyd, Edinburgh, pp 185-210

Borley GD (1967) Potassium-rich volcanic rocks from southern Spain. Mineral Mag 36: 364-379

Bowen NL (1912) The binary system $Na_2Al_2Si_2O_8$ (nepheline, carnegieite)-$CaAl_2Si_2O_8$ (anorthite). Am J Sci 33: 551-573

Bowen NL (1913) The melting phenomena of plagioclase feldspar. Am J Sci 35: 577-599

Bowen NL (1922a) Genetic features of alnoite rocks at Isle Cadieux, Quebec. Am J Sci 3: 1-34

Bowen NL (1922b) The behavior of inclusions in igneous magmas. J Geol 30: 513-570

Bowen NL (1928) The evolution of the igneous rocks. Princeton University Press, New Jersey, p 332

Bowen NL (1937) Recent high temperature research on silicates and its significance in igneous geology. Am J Sci 33: 1-21

Bowen NL, Ellestad RB (1937) Leucite and pseudoleucite. Am Mineral 22: 409-415

Bowen NL, Schairer JF (1935) The system $MgO-FeO-SiO_2$. Am J Sci 29: 151-217

Bragg L, Claringbull GF, Taylor WH (1965) Crystal structures of minerals. G. Bell and Sons, London, p 409

Brown FH (1971) Volcanic petrology of Toro-Ankole region, Western Uganda. Ph D Thesis University of California Berkeley

Brown FH, Carmichael ISE (1969) Quaternary volcanoes of the Lake Rudolf region: Part 1, The basanite tephrite series of the Korath Range. Lithos 2: 239-260

Browne WR (1933) Presidential address, 3 May, 1933. J Proc R Soc NSW 67: 1-95

Bryan WB (1967) Geology and petrology of Calorion Island. Bull Geol Soc Am 78: 1461-1476

Bryan WB, Stice GD, Ewart E (1972) Geology and petrography of Tonga. J Geophys Res 77: 1566-1585

Bucklund HG (1933) On the mode of intrusion of deep seated alkaline rocks. Bull Geol Inst Upsala 24: 1-24

Buddington AF (1922) On some natural and synthetic melilites. Am J Sci Ser 5 3: 35-87

Bullard EC (1936) Gravity measurements of east Africa. Philos Trans R Soc London 235: 445-531

Burnham CW, Holloway JR, Davies NF (1969) Thermodynamic properties of water to 1000 °C and 10,000 bars. Geol Soc Am Spec Pap 132: 96

Caraballo JM (1975) Geoquimica de las rocas lamproticas espanolas. Unpublished Doctoral Thesis. Faculted de Ciencias Geologicas, Universited de Madrid

Carlson HD (1957) Origin of the corundum deposits of Renfrew County, Ontario, Canada. Bull Geol Soc Am 68: 1605-1636

Carman JH (1969) The study of the system $NaAlSiO_4$-Mg_2SiO_4-SiO_2-H_2O from 200 to 5000 bars and 800 °C to 1100 °C and its petrologic applications. Ph D Thesis Pennsylvania State University

Carmichael ISE (1967) The mineralogy and petrology of the volcanic rocks from the Leucite Hills, Wyoming. Contrib Mineral Petrol 15: 24-66

Carmichael ISE, Nicholls J, Smith AL (1970) Silica activity in igneous rocks. Am Mineral 55: 246-263

Carswell DA (1968) Picrite magma-residual dunite relationship in garnet picrite at Kalskaret near Tafjord, South Norway. Contrib Mineral Petrol 19: 97-124

Chatterjee SC (1974) Petrography of the igneous and metamorphic rocks of India. Macmilan Co Ltd, Bombay, p 391

Cloos H (1939) Hebung – Spaltung – Vulkanismus. Geol Rundsch 30: 401-527

Collerson KD, Malpas J (1977) Partial melts in upper mantle nodules from Labrador kimberlites. Extended Abstracts. 2 nd kimberlite Conference Santa Fe

Combe AD (1933) Recent volcanic area of Bunyaruguru. Ann Rep Geol Surv Uganda 1932: 35-36

Combe AD, Holmes A (1945) The kalsilite-bearing lavas of Kabirenge and Lyakauli, South-West Uganda. Trans R Soc Edin 61: 359-379

Coombs DS, Wilkinson JFG (1969) Lineages and fractionation trends in undersaturated volcanic rocks from the East Otago Volcanic Province (New Zealand) and related rocks. J Petrol 10: 440-501

Cross W (1897) The igneous rocks of the Leucite Hills and Pilot Butte, Wyoming. Am J Sci 4: 115-141

Cross W, Iddings JP, Pirsson LV, Washington HS (1902) A quantitative chemico-minerological classification and nomenclature of igneous rocks. J Geol 10: 555-690

Cruft EF (1966) Minor elements in igneous and metamorphic apatite. Geochim Cosmochim Acta 30: 375-398

Cundari A (1973) Petrology of the leucite-bearing lavas in New South Wales. J Geol Soc Aust 20: 465-492

Cundari A (1975) Mineral chemistry and petrogenetic aspects of the Vico lavas, Roman volcanic region Italy. Contrib Mineral Petrol 53: 129-144

Cundari A, Le Maitre RW (1970) On the petrogeny of leucite-bearing rocks of the Roman regions. J Petrol 11: 33-47

Cundari A, Mattias PP (1974) Evolution of the Vico lavas, Roman volcanic region, Italy. Bull Volcanol 38: 98-114

Curran JM (1891) A contribution to the microscopic structure of some Australian rocks. Proc R Soc NSW 25: 179-233

Daly RA (1910) Origin of alkaline rocks. Bull Geol Soc Am 21: 87-118

Daly RA (1933) Igneous rocks and the depths of the earth. McGraw-Hill, New York, p 598

Dawson JB (1966) Oldoinyolengai – an active volcano with sodium carbonatite lava flows. In: Tuttle OF, Gittins J (eds) Carbonatites. John Wiley and Sons, London New York Sydney Toronto, pp 155-168

Dawson JB (1968) Recent researches on kimberlites and diamond geology. Econ Geol 63:504-511

Dawson JB, Smith JV (1977) Late stage diopsides in kimberlitic groundmass. Extended Abstracts. 2nd Int Kimberlite Conf Santa Fe

Deer WA, Howie RA, Zussman J (1963) Rock forming minerals, vol IV. Longmans, London, p 435

Dempster AN, Tucker R (1973) The geology of Sekameng (Butha-Buthe) kimberlite pipe and the associated dyke swarm. In: Nixon PH (ed) Lesotho Kimberlites. Lesotho National Development Corporation, Maseru, pp 180-189

Dickinson WR (1971) Plate tectonic models of geosynclines. Earth Planet. Sci Lett 10:165-174

Discendenti A, Nicoletti M, Taddeucci A (1970) Dalazione K-Ar, e^{230} Th di alcuni produtti del vulcano di Latera (Monti Vulsini). Period Mineral 39:461-468

Donnay G, Schairer JF, Donnay JDH (1959) Nepheline solid solutions. Mineral Mag 32:93-109

Duda A (1975) Petrologie der Basanit Tephrit Reihe des Laacher Seegebietes. Diplomarbeit Ruhr Universität, Bochum

Duda A, Schminke HU (1978) Petrology and chemistry of potassic rocks from the Laacher See area. Neues Jahrb 132 (pt. 1):1-33

Edgar AD (1965) Lattice parameters of melilite solid solutions and a reconnaissance of phase relations in the system $Ca_2Al_2SiO_7$(gehlenite)-$Ca_2MgSi_2O_7$(akermanite)$NaCaAlSi_2O_7$(soda-melilite) at 1000 kg/cm^2 water vapor pressure. Can J Earth Sci 2:596-621

Edgar AD, Green DH, Hibberson WO (1976) Experimental petrology of a highly potassic magma. J Petrol 17:339-356

Egoroff B (1965) L' eruption du volcan Mihaga en 1954. Inst. des. Parcs Nationaux du Congo, Mission d' Etudes Volcanologiques. Fasc 4:91-113

Ehrenberg SN (1977) The Washington Pass volcanic center: Evolution and eruption of minette magmas of the Navajo volcanic field. Extended Abstracts. 2nd Int Kimberlite Conf Santa-Fe New Mexico

Erdley AJ (1961) Structural geology of North America, 2nd edn. Harper and Row, New York, p 624

Erlank AJ, Finger LW (1970) The occurrence of potash richterite in mica nodule from the Wesselton kimberlite, South Africa. Carnegie Inst Wash Yearb 68:442-443

Eugster HP, Wones DR (1962) Stability relations of the ferruginous biotite, annite. J Petrol 3:82-125

Evernden JF, Curtis JF (1965) Potassium-Argon dating of late Cenozoic rocks in East Africa and Italy. Curr Anthropol 6:343-385

Ewart A, Mateen A, Rose JA (1976) Review of mineralogy and chemistry of Tertiary central volcanic complexes in southeast Queensland and northeast New South Wales. In: Johnson RW (ed) Volcanism in Australasia. Elsevier Sci Publ Co, Amsterdam, pp 21-39

Faure G, Powell JL (1975) Strontium isotope geology. Springer, Berlin Heidelberg New York, p 188

Faust GT (1963) Phase transition in synthetic and natural leucite. Schweiz Mineral Petrogr Mitt 43:165-195

Ferguson AK, Cundari A (1975) Petrological aspects of evolution of the leucite-bearing lavas from Bufumbira, Southwest Uganda. Contrib Mineral Petrol 50:25-46

Ferguson JB, Merwin HE (1919) The ternary system $CaO-MgO-SiO_2$. Am J Sci 48:81-123

Finckh L (1912) Die jung-vulkanischen Gesteine des Kiwusee-Gebietes. Wiss Ergeb Dtsch Zentral-Africa Expedition, 1907-1908. 1: 44-58

Fisher JR, Zen E-an (1971) Thermochemical calculations from hydrothermal phase equilibrium data and the free energy of formation of H_2O. Am J Sci 270: 299-314

Forbes WC, Flower MFJ (1974) Phase relations of titanphlogopite, $K_2Mg_4TiAl_2Si_6O_{20}(OH)_4$: A refractory phase in the upper mantle? Earth Planet Sci Lett 22: 60-66

Fornaseri M, Scherillo A, Ventriglia U (1963) La regione vulcanica dei Colli Albani. Consiglio Nazionale delle Recerche, Roma, p 561

Frechen J (1971) Siebengebirge am Rhein, Laacher Vulkangebiet, Maargebiet der Westeifel – Vulcanologisch-Petrographische Exkursionen. Samml Geol Führer 56

Fudali RF (1957) On the origin of pseudoleucite (Abstract). Am Geophys Union Trans 38: 391

Fudali RF (1963) Experimental studies on the origin of pseudoleucite and associated problems of alkalic rock systems. Geol Soc Am Bull 74: 1101-1126

Fyfe WS (1973) Dehydration reactions. Am Assoc Petrogr Geol Bull 57: 190-197

Fyfe WS, McBirney AR (1975) Subduction and the structure of andesitic volcanic belts. Am J Sci 275 A: 285-297

Garlick GD (1966) Oxygen isotope fractionation in igneous rocks. Earth and Planet Sci Letters 1: 301-306

Gass IG (1970) Tectonic and magmatic evolution of the Afro-Arabian dome in African magmatism and tectonics. In: Clifford TN, Gass IG (eds) African magmatism and tectonics. Oliver and Boyd, Edinburgh, pp 285-330

Gast PW (1968) Trace element fractionation and the origin of tholeiite and alkaline magma types. Geochim Cosmochim Acta 32: 1057-1086

Gill JB (1970) Geochemistry of Viti Levu, Fiji and its evolution in an island arc. Contrib Mineral Petrol 27: 179-203

Gittins J (1966) Summaries and bibliographies of carbonatite complex. In: Tuttle OF, Gittins J (eds) Carbonatites. John Wiley and Sons, New York, pp 417-541

Gorai M (1940) A consideration of the genesis of alkali basalts from Wu-ta-lien-chih volcano, North Manchuria. J Geol Soc Jpn 47: 457-467, 481-498

Green DH (1964) The petrogenesis of the high-temperature peridotite intrusion in the Lizard area, Cornwall. J Petrol 5: 134-188

Green DH, Ringwood AE (1963) Mineral assemblages in a model mantle composition. J Geophys Res 68: 937-945

Green DH, Ringwood AE (1967) An experimental investigation of the gabbro to eclogite transformation and its petrological applications. Geochim Cosmochim Acta 31: 767-833

Greig JW, Barth TFW (1938) The system $Na_2O \cdot Al_2O_3 \cdot 2\ SiO_2$ (nepheline, carnegie)-$Na_2O \cdot Al_2O_3 \cdot 6\ SiO_2$ (albite). Am J Sci 5 th Ser 35 A: 93-112

Griffin RJ (1957) The geology and mineral resources of the Eubalong 4-mile military sheet. Tech Rep Dep Mineral N S W 5: 163-190

Griffin WL, Murthy VR (1969) Distribution of K, Rb, Sr, and Ba in some minerals relevant to basalt genesis. Geochim Cosmochim Acta 33: 1389-1414

Gummer WR (1943) The system $CaSiO_3$-$CaAl_2Si_2O_8$-$NaAlSiO_4$. J Geol 51: 503-530

Gupta AK (1972) The system forsterite-diopside-akermanite-leucite and its significance in the origin of potassium-rich mafic and ultramafic volcanic rocks. Am Mineral 57: 1242-1259

Gupta AK, Edgar AD (1974) Phase relations in the system nepheline-leucite-anorthite at 1 atmosphere. Can Mineral 12: 354-356

Gupta AK, Edgar AD (1975) Leucite-Na-Feldspar incompatibility: an experimental study. Mineral Mag 40: 377-384

Gupta AK, Fyfe WS (1975) Leucite survival: The alteration to analcime. Can Mineral 13: 361-363

Gupta AK, Lidiak EG (1973) The system diopside-nepheline-leucite. Contrib Mineral Petrol 41: 231-239

Gupta AK, Yagi K (1977) Experimental study on two picrites with reference to the genesis of kimberlites. 2nd Int Kimberlite Conf Santa Fe, New Mexico. Extend Abstr

Gupta AK, Yagi K (1978) Experimental study on the forsterite-grossularite incompatibility. Abstr Geodynamics and Magma Genesis Symp Tokyo

Gupta AK, Onuma K, Yagi K, Lidiak EG (1973a) Effect of silica concentration on the diopsidic pyroxenes in the system diopside-$CaTiAl_2O_6$-SiO_2. Contrib Mineral Petrol 41: 333-344

Gupta AK, Venkatesawaran GP, Lidiak EG, Edgar AD (1973b) The system diopside-nepheline-akermanite-leucite and its bearing on the genesis of alkali-rich mafic and ultramafic volcanic rocks. J Geol 81: 209-218

Gupta AK, Yagi K, Hariya Y, Onuma K (1976) Experimental investigations of some synthetic leucite rocks under water pressures. Proc Jpn Acad Sci 52: 469-472

Gurney JJ, Ebrahim S (1973) Chemical compositions of Lesotho Kimberlites. In: Nixon PH (ed) Lesotho Kimberlites. Lesotho National Development Corporation, Maseru, pp 280-284

Hamilton DL, MacKenzie WS (1960) Nepheline solid solutions in the system Na-$AlSiO_4$-$KAlSiO_4$-SiO_2. J Petrol 1: 56-72

Hamilton DL, MacKenzie WS (1965) Phase equilibrium studies in the system Na-$AlSiO_4$ (nepheline)-$KAlSiO_4$ (kalsilite)-SiO_2-H_2O. Mineral Mag 34: 214-231

Harker RI, Tuttle OF (1956) The lower limit of stability of akermanite (Ca_2Mg-Si_2O_7). Am J Sci 254: 468-474

Harris PG (1957) Zone refining and origin of potassic melts. Geochim Cosmochim Acta 12: 195-208

Harris PG (1969) Basalt type and rift valley tectonism. Tectonophysics 8: 427-436

Harumoto A (1930) Soda sanidinite from Utsuryoto island. Commemoration volume on the occasion of Professor Ogawa's 60th birthday. Tokyo, pp 539-548

Harumoto A (1970) Volcanic rocks and associated rocks of Utsuryoto Island (Japan Sea). Nippon Printing Publishing Co, Osaka, pp 1-39

Hatherton T, Dickinson WR (1968) Andesite volcanism and seismicity in New Zealand. J Geophys Res 73: 4615-4619

Hatherton T, Dickinson WR (1969) The relationship between andesitic volcanism and seismicity in Indonesia, the Lesser Antilles and other island arcs. J Geophys Res 74: 5301-5310

Heinrich EW (1966) The geology of carbonatites. McNally and Co, Chicago, p 607

Higazy A (1954) Trace elements of volcanic and ultrabasic potassic rocks of southwestern Uganda and adjoining part of Belgian Congo. Geol Soc Am Bull 65: 39-70

Hijikata K, Onuma K (1969) Phase equilibrium of the system $CaMgSi_2O_6$-$CaFe^{3+}$-$AlSiO_6$ in air. J Jpn Assoc Mineral Petrol Econ Geol 62: 209-217

Hoefs J, Wedepohl KH (1968) Strontium isotope studies on young volcanic rocks from Germany and Italy. Contrib Mineral Petrol 19: 328-338

Holmes A (1932) The origin of igneous rocks. Geol Mag 69: 543-558

Holmes A (1937) A contribution to petrology of kimberlite and its inclusion. Trans Geol Soc S Afr 39: 379-428

Holmes A (1942) A suite of volcanic rocks from southwest Uganda containing kalsilite (a polymorph of $KAlSiO_4$). Mineral Mag 26: 197-217

Holmes A (1950) Petrogenesis of katungite and its associates. Am Mineral 35: 772-792

Holmes A (1965) Principles of physical geology, 2nd edn. Ronald Press, New York, p 1069-1078

Holmes A, Harwood HF (1932) Petrology of the volcanic fields east and southeast of Ruwenzori, Uganda. Q J Geol Soc 88: 370-442

Holmes A, Harwood HF (1937) The petrology of the volcanic rocks of Bufumbira. Mem Geol Surv Uganda 3: 1-300

Holmes A, Paneth FA (1936) Helium ratios of rocks and minerals from the diamond pipes of south Africa. Proc R Soc 154A: 385-413

Huckenholz HG, Schairer JF, Yoder HS (1969) Synthesis and stability of ferri diopside. Mineral Soc Am Spec Pap 2: 163-177

Hurley PM, Fairbairn HV, Pinson WH Jr (1966) Rb-Sr isotope evidence in the origin of potash-rich lavas of Western Italy. Earth Planet Sci Lett 1: 301-306

Hussak E (1890) Über Leucit-Pseudokrystalle in Phonolith (Tinguait) der Serre de Tingua, Estado Rio de Janeiro, Brazil. Neues Jahrb 1: 166-169

Hytönen K, Schairer JF (1961) Crystallization in forsterite-anorthite-diopside-silica. Carnegie Inst Wash Yearb 60: 139-141

Iddings JP (1913) Igneous rocks. John Wiley and Sons Inc, London, p 685

Iddings JP, Morley EW (1915) Contribution to the petrography of Java and Celebes. J Geol 23: 231-245

Ikeda K, Yagi K (1972) Synthesis of kosmochlor and phase equilibria in the join $CaMgSi_2O_6-NaCrSi_2O_6$. Contrib Mineral Petrol 36: 63-72

Imbo G (1965) Catalogue of active volcanoes of the world including solfatara fields. Int Assoc Volcanol 18: 1-71

Irving AJ (1974) Cr-diopside in olivine-rich lherzolite nodules. Neues Jahrb Mineral Abh 120: 147-167

Jakes P, White AJR (1969) Structure of Melanesium arcs and correlation with distribution of magma types. Earth Planet Sci Lett 8: 223-236

Jakes P, White AJR (1971) Composition of Island arc and continental growth. Earth Planet Sci Lett 12: 224-230

Johansen A (1931) A descriptive petrography of igneous rocks. Chicago University Press, Chicago, p 239, 280

Judd JW (1887) On the discovery of leucite in Australia. Mineral Mag 7: 194-195

Kelley KK (1960) Contributions to the data on theoretical metallurgy, part 13, High temperature heat content, heat capacity and entropy data for the elements and inorganic compounds. US Bureau Mines Bull 584: 232

Kemp JF (1897) The Leucite Hills of Wyoming. Bull Geol Soc Am 8: 169-182

Kemp JF, Knight WC (1903) Leucite Hills of Wyoming. Bull Geol Soc Am 14: 305-336

Kennedy GC (1953) Geology and mineral deposits of Jumbo basin, southeastern Alaska. U S G S Prof Pap 251: 1-46

Kennedy GC (1955) Some aspects of the role of water in rock melts. Geol Soc Am Spec Pap 62: 489-503

King BC (1949) The Napak area of southern Karamoja. Geol Surv Uganda Mem 5: 57

King BC (1955) The Ard Bheinn area of the central igneous complex of Arran. Q J Geol Soc London 110: 323-356

King BC (1965) Petrogenesis of the alkaline igneous rock suites of the volcanic and intrusive centers of Eastern Uganda. J Petrol 6: 67-100

King BC, Sutherland DS (1966) The carbonatite complexes of Eastern Uganda. In: Tuttle OF, Gittins J (eds) Carbonatites. John Wiley and Sons, New York, pp 73-126

Knight CW (1906) A new occurrence of pseudoleucite. Am J Sci 21:286-293

Kresten P, Dempster AN (1973) The geology of pipe 200 and the Malibamatso dyke swarm. In: Nixon PH (ed) Lesotho kimberlites. Lesotho National Development Corporation, Maseru, pp 172-179

Koster van Groos AF, Wyllie PJ (1966) Liquid immisibility in the system $Na_2O-Al_2O_3-SiO_2-CO_2$ at pressures to 1 kilobar. Am J Sci 264:234-255

Koster van Groos AF, Wyllie PJ (1968) Liquid immiscibility in the system $NaAlSi_3O_8-Na_2CO_3-H_2O$ and its bearing on the genesis of carbonatites. Am J Sci 266:932-937

Koster van Groos AF, Wyllie PJ (1973) Liquid immiscibility in the join $NaAlSi_3O_8-CaAl_2Si_2O_8-Na_2CO_3-H_2O$. Am J Sci 273:465-487

Kuno H (1959) Origin of Cenozoic petrographic provinces of Japan and surrounding areas. Bull Vulcanol Ser 2 20:37-76

Kushiro I, Erlank AJ (1970) Stability of potassic richterite. Carnegie Inst Wash Yearb 68:231-233

Kushiro I, Yoder HS (1966) Anorthite-forsterite and anorthite-enstatite reactions and their bearing on the basalt-eclogite transformation. J Petrol 7:337-362

Lacroix A (1917) Sur la transformation de quelques roches eruptive besiques en amphibolites. C R Acad Sci 164:969-974

Larimer WM (1964) The oxidation states of the elements and their potentials in aqueous solutions. Englewood Cliffs. NJ Prentice Hall Inc: p 392

Larsen ES (1940) Petrographic province of central Montana. Geol Soc Am Bull 51:887-945

Larsen ES, Buie BF (1938) Potash analcite and pseudoleucite from the Highwood Mountains of Montana. Am Mineral 23:837-849

Larsen ES, Buie BF (1941) Igneous rocks of the Highwood Mountains, Part 5. Geol Soc Am Bull 52:1829-1840

Larsen ES, Hurlbut CS, Burgess CH, Buie BF (1941) Igneous rocks of Highwood Mountains. Geol Soc Am Bull 52:1733-1868

Le Bas MJ (1977) Carbonatite-nephelinite volcanism. John Wiley and Sons, London New York Sydney Toronto, p 348

Le Maitre RW (1962) Petrology of volcanic rocks, Gough Island, South Atlantic. Bull Geol Soc Am 73:1309-1342

Le Maitre RW (1968) Chemical variation within and between volcanic rock-series – a statistical approach. J Petrol 9:220-252

Le Maitre RW (1969) Kaersutite-bearing plutonic xenoliths from Tristan da Cunha, South Atlantic. Mineral Mag 37:185-197

Le Pichon X (1968) Sea-floor spreading and continental drift. J Geophys Res 73:3661-3697

Lewis HC (1897) Bonney TG (ed) Papers and Notes on the genesis and the matrix of diamond. Longmans Green and Company, London, p 72

Lindgren W (1933) Differentiation and ore deposition: Lindgren volume: Ore deposits of the Western States. Am Inst Min Metall Eng 797:152-180

Lindsley DH (1966) P-T projection for part of the system kalsilite-silica. Carnegie Inst Wash Yearb 65:244-247

Lloyd FE, Bailey DK (1969) Carbonatites in the tuffs of W Eifel area of Germany. Contrib Mineral Petrol 23:136-139

Locardi E, Sircana S (1967) Distributione del Lazio settentrionale. R Soc Mineral Ital 23:163-224

Luth WC (1967) Studies in the system $KAlSiO_4-Mg_2SiO_4-SiO_2-H_2O$: I. Inferred phase relations and applications. J Petrol 8:372-416

Luth WC (1974) Resorption of silicate minerals. In: Sorensen H (ed) The alkaline rocks. John Wiley and Sons, London New York Sydney Toronto, pp 500-515

Macdonald GA, Katsura T (1964) Chemical composition of Hawaiian lavas. J Petrol 5:82-133

MacKenzie WS, Rahman S (1968) The paragenesis of leucite-Na-Feldspar. Contrib Mineral Petrol 19:339-342

Marinelli G, Mittempergher M (1966) On the genesis of some magmas of typical Mediterranean (potassic) suite. Bull Volcanol 29:113-140

Markov VK, Petrov VP, Delisin IS, Rayabinin YN (1966) Phlogopite transformations at high pressures and temperatures. Geochem Int 2:1112-1120

Markov VK, Petrov VP, Delsin IS, Rayabinin YN (1968) Transformations of biotite and lepidomelane at high pressures and temperatures. Int Geol Rev 10:1028-1036

Mattias PP (1965) Lada dell'apparato Vulsino. Period Mineral 34:137-199

Mattias PP, Ventriglia U (1970) La regione vulcanica dei Monti Sabatine Cimini. Mem Soc Geol Ital 9:331-384

McBirney AR (1969) Compositional variations in Cenozoic calc alkaline suites of Central America. State Oregon Dept Geol Mineral Ind Bull 65:185-189

McBirney AR, Aoki K (1968) Petrology of the Island of Tahiti. Geol Soc Am. Mem 116:523-556

Merrill RB, Wyllie PJ (1975) Kaersutite and kaersutite eclogite melting from Kakanui, New Zealand – water-excess and water deficient melting to 30 kilobars. Geol Soc Am Bull 86:555-570

Meyer A (1953) Les basaltes du Kivu. Meriditional Serv Geol Congo Belge Ruanda-Urundi Mem No 2:231-242

Miyashiro A (1951) The range of chemical composition in nepheline and their petrogenetic significance. Geochim Cosmochim Acta 1:278-283

Miyashiro A (1960) Thermodynamics of reactions of rock-forming minerals with silica, Part 1. Jpn J Geol Geogr 33:71-78

Miyashiro A (1975) Island arc volcanic series: a critical review. Petrologie 1:177-182

Modreski PJ, Boettcher AL (1972) The stability of phlogopite enstatite at high pressures: A model for micas in the interior of the earth. Am J Sci 272:852-869

Modreski PJ, Boettcher AL (1973) Phase relationships of phlogopite in the system $K_2O-MgO-CaO-Al_2O_3-SiO_2-H_2O$ to 35 kilobars: A better model for micas in the interior of the earth. Am J Sci 273:385-414

Morey GW, Fleischer M (1940) Equilibrium between vapour and liquid phases in the system $H_2O-CO_2-K_2O-SiO_2$. Bull Geol Soc Am 51:1035-1058

Morey GW, Hesselgesser JM (1952) The system $H_2O-Na_2O-SiO_2$ at 400 °C. Am J Sci Bowen Vol:343-371

Morgan HJ (1972) Deep mantle convection plume and plate motions. Am Assoc Petrol Geol Bull 56:203-213

Morse SA (1969) Alkali feldspar-water at 5 kb. Carnegie Inst Wash Yearb 67:120-126

Moyd L (1949) Petrology of nepheline and corundum rocks of south eastern Ontario. Am Mineral 34:736-751

Mügge C (1927) Zur Kenntnis des Kaliophilit. Z Kristall 65:380-387

Muir ID, Tilley CE (1961) Mugearites and their place in alkali igneous rock series. J Geol 69:186-203

Mysen BO, Boettcher AL (1975) Melting of hydrous mantle: II. Geochemistry of crystals and liquids formed by anatexis of mantle peridotite at high pressures and high temperatures as a function of controlled activities of water, hydrogen, and carbon dioxide. J Petrol 16: 520-548

Nakamura Y, Yoder HS (1974) Analcite, hyalophane, and phillipsite from the Highwood Mountains, Montana. Carnegie Inst Wash Yearb 73: 354-358

Nappi G (1969) Stratigrafiae petrographia dei Vulsini Sud Occidentali (Caldera di Latera). Boll Soc Geol Ital 88: 171-181

Nash WP, Wilkinson JFG (1970) Shonkin Sag Laccolith Montana: part 1, Mafic minerals and estimates of pressure, oxygen fugacity and silica activity. Contrib Mineral Petrol 25: 241-269

Nash WP, Wilkinson JFG (1971) Shonkin Sag Laccolith Montana: Pt. II, Bulk rock geochemistry. Contrib Mineral Petrol 33: 162-170

Ninkovich D, Hays JD (1972) Mediterranean island arcs and origin of potash volcanoes. Earth Planet Sci Lett 16: 331-345

Nixon PH, Boyd FR (1973a) Deep seated nodules. In: Nixon PH (ed) Lesotho kimberlites. Lesotho National Development Corporation, Maseru, pp 106-109

Nixon PH, Boyd FR (1973b) Notes on the heavy mineral concentrates. In: Nixon PH (ed) Lesotho kimberlite. Lesotho National Development Corporation, Maseru, pp 218-220

Ogura T, Matsumoto M (1938) Report on the geology of Erh-k'o Shan Volcano, Lungchiang Province. Report on the volcanoes of Manchuria, No 2. Ryojun Coll Eng

Ogura T, Matsuda K, Nakagawa T, Matsumoto M, Murata K (1936) Report on the geology of Wu-ta-lien-chih Volcano, Lungchiang Province. Report on the volcanoes of Manchuria, No 1. Ryojun Coll Eng p 96

Ogura T, Sawatari M, Murayama K (1939) Report on the geology of Ch'i-hsing Shan Volcano, Feng-t'ien Province and Hsing-an Province. Report on the volcanoes of Manchuria, No 3. Ryojun Coll Eng p 44

O'Hara MJ (1965) Primary magma and the origin of basalts. Scott J Geol 1: 19-40

O'Hara MJ, Yoder HS (1967) Formation and fractionation of basic magmas at high pressures. Scott J Geol 3: 67-117

Onuma K, Yagi K (1967) The system diopside-akermanite-nepheline. Am Mineral 52: 227-243

Onuma K, Iwai T, Yagi K (1972) Nepheline-iron nepheline solid solutions. J Fac Sci Hokkaido Univ Ser 4 15: 179-190

Orville PM (196) Unit-Cell parameters of the microcline-low albite and the sanidine-high albite solid solution series. Am Mineral 52: 58-86

Osann A (1906) Über einige Alkaligesteine aus Spanien. Festschr. Rosenbusch Stuttgart 263-310

Osborn EF, Schairer JF (1941) The ternary system pseudo-wollastonite-akermanite-gehlenite. Am J Sci 239: 715-733

Osborn EF, Tait DB (1952) The system diopside-forsterite-anorthite. Am J Sci Bowen Vol 2: 413-433

Oxburgh ER (1964) Petrological evidence for the presence of amphibole in the upper mantle and its petrogenetic and geophysical implications. Geol Mag 101: 1-19

Oxburgh ER, Turcotte DL (1970) Thermal structure of Island Arcs. Bull Geol Soc Am 81: 1665-1688

Padang MN Van (1951) Catalogue of the active volcanoes of Indonesia. Int Volcanol Assoc 1: 1-271

Patterson EM (1951) A petrochemical study of the Tertiary Lavas of north-east Ireland. Geochim Cosmochim Acta 2: 283-299

Pearce TH (1970) The analcite-bearing volcanic rocks of the Crowsnest formation, Alberta. Can J Earth Sci 7: 46-66

Pecora WT (1942) Nepheline syenite pegmatites Rockey Boy stock, Bearpaw Mountains, Montana. Am Mineral 27: 397-424

Perchuk LL, Ryabchikov ID (1968) Mineral equilibria in the system nepheline-alkali feldspar-plagioclase and their petrological significance. J Petrol 9: 123-167

Peterschmidt E (1957) Quelques donnes nouvelles sur les sesimes profonds de la Mer Tyrrhenienne. Anal Geofis 9: 305-318

Platt RG, Edgar AD (1971) The system nepheline-diopside-sanidine and its significance in the genesis of melilite and olivine-bearing alkaline rocks. J Geol 80: 224-236

Poldervaart A, Hess HH (1951) Pyroxenes in the crystallization of basaltic magma. J Geol 59: 472-489

Powell JL, Bell K (1970) Strontium isotopic studies of alkalic rocks. Localities from Australia, Spain, and the Western United States. Contrib Mineral Petrol 27: 1-10

Prider RT (1939) Some minerals from the leucite-rich rocks of West Kimberley area of Western Australia. Mineral Mag 25: 373-383

Prider RT (1960) The leucite lamproites of Fitzroy basin, Western Australia. J Geol Soc Aust 6: 71-118

Rahman S (1975) Some aluminous clinopyroxenes from Vesuvius and Monte Somma, Italy. Mineral Mag 40: 43-52

Raleigh CB (1967) Tectonic implications of serpentinitic weakening. Geophys J R Astron Soc 14: 113

Rath G vom (1864) Skizzen aus dem vulkanischen Gebiet des Niederrheins. 2: Fortsetz Z D Geol Ges 16: 82

Ricker RW, Osborn EF (1954) Additional phase diagram of the system CaO-MgO-SiO₂. Am Ceram Soc J 37: 133-139

Ringwood AE, Kesson SE (1977) Synthesis of pyrope-knorringite garnet paragenesis. Extended abstracts. 2nd Int Kimberlite Conf Santa Fe, New Mexico

Rittmann A (1931) Gesteine und Mineralien von Monte Vulture in der Basilicata. Schweiz Mineral Petrol Mitt 11: 240-252

Rittmann A (1933) Die geologisch bedingte Evolution und Differentiation des Somma-Vesuvius magma. Z Vulkanol 5: 8-94

Rittmann A (1951) Magmatic character and tectonic position of the Indonesian Volcanoes. Nomenclature of volcanic rocks. Bull Volcanol Ser II Tome 12: 46-58

Robie RA, Waldbaum DR (1968) Thermodynamic properties of minerals and related substances at $298.15\,^{\circ}K$ ($25.0\,^{\circ}C$) and one atmosphere (1.013 bars) pressure at higher temperatures. Geol Soc Am Bull 1259: 256

Roedder EW (1951) The system $K_2O-MgO-SiO_2$. Am J Sci 249: 81-130, 224-248

Roedder E (1965) Liquid CO_2 inclusion in olivine-bearing nodules and phenocrysts from basalts. Am Mineral 50: 1746-1782

Rosenqvist I, Th (1951) Investigations in the crystal chemistry of silicates. III. The relation hematite-microline. Nor Geol Tidsskr 29: 65

Rutherford, MJ (1969) An experimental determination of iron biotite-alkali feldspar equilibria. J Petrol 10: 381-408

Sabatini V (1900) I. Vulcani dell'Italia Centrale ei loro prodotti. Parte I. R. Uffico Geol Volcano Laziale Roma

Sahama ThG, Wiik HB (1952) Leucite, potash nepheline and clinopyroxene from volcanic lavas from southwestern Uganda and adjoining Belgian Congo. Am J Sci Bowen Vol: 457-470

Sahama ThG (1953a) Mineralogy and petrology of a lava flow from Mt. Nyiragongo, North Kivu, Belgian Congo. Ann Acad Sci Fenn Ser A-III 35: 1-11

Sahama ThG (1953b) Parallel growths of nepheline and microperthitic kalsilite from North Kivu, Belgian Congo. Ann Acad Sci Fenn Ser A-III 36: 1-15

Sahama ThG (1954) Mineralogy of mafurite. Bull Comm Geol Finl 166: 21-28

Sahama ThG (1957) Complex nepheline-kalsilite phenocryst in Kabfumn lava, Nyiragongo, area, North Kivu in Belgian Congo. J Geol 65: 515-530

Sahama ThG (1960) Kalsilite in the lavas of Mt. Nyiragongo (Belgian Congo). J Petrol 1: 146-171

Sahama ThG (1973) Evolution of Nyiragongo magma. J Petrol 14: 33-48

Sahama ThG, Neuvonen KJ, Hytönen K (1956) Determination of the composition of kalsilites by an X-ray method. Mineral Mag 31: 200-210

Savelli C (1967) The problem of rock assimilation by Somma-Vesuvius magma. Contrib Mineral Petrol 16: 328-353

Scarfe CM, Luth WC, Tuttle OF (1966) An experimental study bearing on the absence of leucite in plutonic rocks. Am Mineral 51: 726-735

Schairer JF (1944) Some aspects of the melting and crystallization of rock forming minerals. Am Mineral 29: 75-95

Schairer JF (1948) Phase equilibrium relations in the quaternary system $K_2O-MgO-Al_2O_3-SiO_2$ (preliminary report) Abstr Geol Soc Am Bull 59: 1349

Schairer JF (1950) The alkali feldspar join in the system $NaAlSiO_4-KAlSiO_4-SiO_2$. J Geol 58: 512-517

Schairer JF (1954) The system $K_2O-MgO-Al_2O_3-SiO_2$: I, Results of quenching experiments on four joins in the tetrahedron cordierite-forsterite-leucite-silica and on the join cordierite-mullite-potash feldspar. Am Ceram Soc J 37: 501-533

Schairer JF, Bowen NL (1935) Preliminary report on the equilibrium relations between feldsparhoids, alkali-feldspars and silica. Trans Am Geophys Union 16th Ann Meet: 325-328

Schairer JF, Bowen NL (1938) The system leucite-diopside-silica. Am J Sci 5th Ser 35A: 289-309

Schairer JF, Bowen NL (1947) The system anorthite-leucite-SiO_2. Geol Finl Bull 20: 67-87

Schairer JF, Bowen NL (1955) The system $K_2O-Al_2O_3-SiO_2$. Am J Sci 253: 681-746

Schairer JF, Yoder HS (1964a) The join akermanite ($Ca_2MgSi_2O_7$)-soda-melilite ($NaCaAlSi_2O_7$). Carnegie Inst Wash Yearb 63: 89-90

Schairer JF, Yoder HS (1964b) Crystals and liquid trends in simplified basalts. Carnegie Inst Wash Yearb 63: 64-74

Schairer JF, Yoder HS (1970) Critical planes and flow sheet for a portion of the system $CaO-MgO-Al_2O_3-SiO_2$. Carnegie Inst Wash Yearb 68: 202-214

Schairer JF, Yagi K, Yoder HS (1962) The system nepheline-diopside. Carnegie Inst Wash Yearb 61: 96-98

Schairer JF, Yoder HS, Tilley CE (1965) Behavior of melilites in the join gehlenite-soda-melilite-akermanite. Carnegie Inst Wash Yearb 64: 95-100

Schairer JF, Yoder HS, Tilley CE (1967) The high temperature behavior of synthetic melilites in the join gehlenite-soda melilite-akermanite. Carnegie Inst Wash Yearb 65: 217-226

Schneider H (1965) Petrographie des Lateravulkans und die Magmenentwicklung der Monti Volsini (Prov. Grosseto, Viterbo und Orvieto, Italien). Schweiz Mineral Petrol Mitt 45: 331-455

Seki Y, Kennedy GC (1964) An experimental study on the leucite-pseudoleucite problem. Am Mineral 49: 1267-1280

Shand SJ (1910) On borolannite and its associates in Assynt. Trans Edinburgh Geol Soc 9: 202-210

Shand SJ (1931) The granite-syenite-limestone complex of Palabora, Transvaal. Trans Geol Soc S Afr 67: 81-92

Shand SJ (1934) The heavy minerals of kimberlite. Trans Geol Soc S Afr 37: 57-68

Shand SJ (1939) Loch Borolan Laccolith, Northwest Scotland. J Geol 47: 408-420

Shand SJ (1943) Eruptive rocks, 2nd edn. John Wiley and Sohns, New York, p 444

Simkin T, Smith JV (1970) Minor element distribution in olivine. J Geol 78: 304-325

Smith JV, Tuttle OF (1957) The nepheline-kalsilite system: I, X-ray data for the crystalline phases. Am J Sci 255: 282-305

Smith JV, Yoder HS (1956) Experimental and theoretical studies of the mica polymorphs. Mineral Mag 31: 209-235

Sobolev VS, Bazarova TJ, Yagi K (1975a) Crystallization temperature of wyomingite from Leucite Hills. Contrib Mineral Petrol 49: 301-308

Sobolev VS, Sobolev NV, Lavrent'ev YuG (1975b) Chrome-rich clinopyroxenes from the kimberlites of Yakutia. Neues Jahrb Mineral Abh 123: 213-218

Stonier GA (1893) On the occurrence of leucite basalt at Lake Cudagellico (Cargellico). Rec Geol Surv N S W 3: 71-74

Stewart FH (1941) On sulphatic cancrinite and analcime (eudnophite) from Loch Borolan Assynt. Mineral Mag 26: 1-5

Sugimura A (1960) Zonal arrangement of geophysical and petrological feature in Japan and its environs. Tokyo Univ Fac Sci J Ser 2 12: 133-153

Takeshita H (1974) Petrological studies on the volcanic rocks of the northern Fossa Magna region, Central Japan, part 1. Pac Geol 7: 65-92

Takeshita H (1975) Petrological studies of volcanic rocks of the northern Fossa Magna region, Central Japan, part 2. Pac Geol 10: 1-32

Taljaard MS (1936) South African melilite basalts and their relations. Trans Geol Soc Afr 39: 281-316

Tateiwa I (1976) The Koreo-Japanese geotectonic zone. Tokyo University Press, Tokyo, p 654

Taylor D, MacKenzie WS (1975) A contribution to the pseudoleucite problem. Contrib Mineral Petrol 49: 321-333

Taylor HP, Epstein S (1963) O^{18}/O^{16} ratios in rocks and coexisting minerals of the Skaergaard intrusion, East Greenland. J Petrol 4: 51-74

Tazieff H (1966) Etat actuel des connassaisance sur le volcan Niragongo (Republique Democratique du Congo). Bull Soc Geol Fr 8: 176-200

Templeman-Kluit DJ (1969) Reexamination of pseudoleucite from the Spotted Fawn Creek, West Central Yukon. Can J Earth Sci 6: 55-62

Thompson RN (1972) Oscillatory and sector zoning in augite from Vesuvian lava. Carnegie Inst Wash Yearb 71: 463-470

Tilley CE (1954) Nepheline-alkali feldspar paragenesis. Am J Sci 252: 65-75

Tilley CE (1958) The leucite nepheline dolerites of Meiches, Vogelsberg, Hessen. Am Mineral 43: 758-761

Tilley CE, Henry NFM (1953) Latiumite (sulphatic potassium-calcium-aluminum silicate) a new mineral from Albano Latium, Italy. Mineral Mag 30: 39-45

Tilley CE, Muir ID (1962) The Hebridean magma type, Edinburgh. Geol Soc Trans 19: 208-215

Tilley CE, Muir ID (1964) Intermediate members of the oceanic basalt-trachyte association. Geol Foeren Stockholm Foerh 85: 436-444

Tilley CE, Yoder HS, Schairer JF (1965) Melting relations of volcanic tholeiite and alkali rock series. Carnegie Inst Wash Yearb 64: 69-82

Toksöz MN, Minear JW, Julian BR (1972) Temperature field and geophysical effects of a downgoing slab. J Geophys Res 76: 1113-1138

Tomilson WH (1939) Corundum in dike at Glenn Riddle Pennsylvania. Am Mineral 24: 339-343

Tomisaka T, Eugster HP (1968) Synthesis on the sodalite group and subsolidus equilibria in the sodalite-nosean system. Mineral J 5: 249-275

Tomita T (1967) Volcanic geology of the Cenozoic alkaline petrographic province of eastern Asia. Geol Mineral Resources Far East 1: 139-202

Trigila R (1969) Sulla genesi dei magmi a carattere mediterraneo. Nota I. II comportmento dei litotipi del settore vulkanico di Latera conriferimento al modello di differenziazione per cristallizazione frazionata. Period Mineral 38: 625-660

Tsuboi S (1920) On a leucite rock vulsinite vicoite from Utsuryoto Island in the sea of Japan. Geol Soc Jpn 27: 91-103

Turcotte DL, Oxburgh ER (1972) Mantle convection and the new global tectonics. Ann Rev Fluid Mech 4: 33-68

Turi B, Taylor HP (1976) Oxygen Isotope studies of potassic volcanic rocks of the Roman Province, central Italy. Contrib Mineral Petrol 55: 1-31

Turner FJ, Verhoogen J (1960) Igneous and metamorphic petrology. McGraw-Hill Co, New York London, p 694

Tuttle OF, Bowen NL (1958) Origin of granite in the light of experimental studies in the system $NaAlSi_3O_8-KAlSi_3O_8-SiO_2-H_2O$. Geol Soc Am Mem 74: p 153

Tuttle OF, Gittins J (1966) Carbonatites. John Wiley and Sons, New York, p 541

Tuttle OF, Smith JV (1958) The nepheline-kalsilite system II: Phase relations. Am J Sci 256: 571-589

Ukhanov AV (1963) Olivine melilitite from a diamond-bearing pipe at Anabar. An SSSR 153: 6 (Translated in Diklady Adak Nauk SSR 153: 176-180)

Varne R (1968) The petrology of Moroto Mountain, Eastern Uganda, and the origin of nephelinites. J Petrol 9: 169-190

Venkateswaran GP (1973) The system nepheline-akermanite-leucite. M Sc Thesis University of Western Ontario, London Canada

Volmer R (1976) Sr and Pb isotope studies of potassic volcanic rocks of the Roman Province, Italy. Rb-Sr and U-Th-Pb systematic alkalic rocks from Italy. Geochim Cosmochim Acta 40: 283-290

Wade A, Prider RT (1940) The leucite-bearing rocks in the West Kimberley area, Western Australia. Geol Soc London Q J 96: 39-98

Wagner PA (1914) The diamond fields of southern Africa. The Transvaal Leader, Johannesburg 2nd impression Cape Town: Struik (PTY) Ltd p 314

Walter LS (1963) Experimental studies on Bowen's decarbonation series, I, P-T univariant equilibria of the monticellite and "akermanite" reactions. Am J Sci 261: 488-500

Warner RD, Luth WC (1973) Two phase data for the join monticellite ($CaMg-SiO_4$)-forsterite (Mg_2SiO_4): experimental study and numerical analyses. Am Mineral 58: 998-1008

Washington HS (1906) The Roman Comagmatic Region. Carnegie Inst Wash Pub No 57: pp 140

Washington HS (1923) Italite, a new rock. Am J Sci Ser 5 1: 33-42

Washington HS (1927) Italite locality of Vila Senni. Am J Sci 14: 173-182

Washington HS, Merwin HE (1923) Augite of the Alban Hills. Am Mineral 8: 104-110

Watkinson DH (1973) Pseudoleucite from plutonic alkalic rock-carbonate complexes. Can Mineral 12: 129-134

Watkinson DH, Wyllie PJ (1969) Phase equilibrium studies bearing on the limestone assimilation hypothesis. Bull Geol Soc Am 80: 1565-1576

Weed WH, Pirsson LV (1896) Missourite, a new leucite rock from Highwood Mountains of Montana. Am J Sci 2: 315-323

White RW (1966) Ultramafic inclusions in basaltic rocks from Hawaii. Contrib Mineral Petrol 12: 245-314

Williams H (1936) Pliocene volcanoes of Navajo-Hopi County. Bull Geol Soc Am 47: 111-172

Willemse J, Bensch JJ (1964) Inclusions of original carbonate rocks in gabbro and norite of the eastern part of the Bushveld Complex. Trans Geol Soc S Afri 67: 1-87

Wimmenauer W (1974) The alkaline province of Central Europe and France. In: Sorensen H (ed) The alkaline rocks. John Wiley and Sons, London New York Sydney Toronto, pp 238-271

Winchell AN, Winchell H (1964) The microscopic character of artificial and inorganic solid substances. John Wiley and Sons, New York, pp 451

Wones DR (1967) A low pressure investigation of the stability of phlogopite. Geochim Cosmochim Acta 31: 2248-2253

Wones DR, Eugster HP (1965) Stability of biotite: experimentals, theory and application. Am Mineral 50: 1228-1272

Wright JB (1963) A note on possible differentiation trends in Tertiary to Recent lavas of Kenya. Geol Mag 100: 164-180

Wyllie PJ (1966) Experimental studies of carbonatite problems: the origin and differentiation of carbonatite magmas. In: Tuttle OF, Gittins J (eds) Carbonatites. John Wiley and Sons, New York, pp 311-352

Wyllie PJ (1970) Ultramafic rocks and the upper mantle. Mineral Soc Am Spec Pap 3: 3-32

Wyllie PJ (1971) The dynamic earth. John Wiley and Sons, New York London Sydney Toronto, p 416

Wyllie PJ (1973) Experimental petrology and global tectonics - a preview. In: Wyllie PJ (ed) Experimental petrology and global tectonics. Tectonophysics 17: 189-203

Wyllie PJ (1974) Limestone assimilation. In: Sorensen H (ed) The alkaline rocks. John Wiley and Sons, London New York Sydney Toronto, pp 459-474

Yagi K (1953) Petrochemical studies on the alkalic rocks of Morotu District, Sakhalin. Geol Soc Am Bull 64: 769-810

Yagi K (1954) Pseudoleucite from Tzu Chin Shan, Shansi, North China. Jpn J Geol Geogr 24: 93-110

Yagi K (1969) Petrology of the alkalic dolerites of the Nemuro Peninsula, Japan. In: Larsen L (ed) Igneous and metamorphic geology. Geol Soc Am Mem 115: 103-147

Yagi K, Gupta AK (1977) Experimental study on assimilation reactions of some granitic rocks related to the genesis of leucite rocks. Bull Volcanol Soc Jpn 22: 65-74

Yagi K, Gupta AK (1978) On the genesis of pseudoleucite with special reference to the pseudoleucite from Tezhsarsk, USSR. Abstract. Int Mineral Assoc Meet Novosibirsk U S S R

Yagi K, Matsumoto H (1966) Note on the leucite-bearing rocks of Leucite Hills, Wyoming, USA. J Fac Sci Hokkaido Univ Ser 4 13: 301-312

Yagi K, Onuma K (1967) The join $CaMgSi_2O_6-CaTiAl_2O_6$ and its bearing on the titanaugites. J Fac Sci Hokkaido Univ Ser 4 13: 463-483

Yagi K, Hariya Y, Onuma K, Fukushima N (1975) Stability relation of kaersutite. J Fac Sci Hokkaido Univ Ser 4 16: 331-342

Yamanari F (1920) Alkalic rocks from Kisshu and Meisen districts, Kan-Kyo-hokudo, Korea. J Geol Soc Tokyo 28: 221-230

Yoder HS (1968) Akermanite and related melilite-bearing assemblages. Carnegie Inst Wash Yearb 66: 471-477

Yoder HS (1969) Anorthite-akermanite and albite-soda melilite reaction relations. Carnegie Inst Wash Yearb 67: 105-108

Yoder HS (1973) Melilite stability and paragenesis. Fortschr Mineral 50: 140-173

Yoder HS (1976) Generation of basalt magma. National Academy of Sciences, Washington DC, p 264

Yoder HS, Kushiro I (1969) Melting of hydrous phase phlogopite. Am J Sci 267: 558-582

Yoder HS, Schairer JF (1969) The melilite-plagioclase incompatibility dilemma in igneous rocks. Carnegie Inst Wash Yearb 67: 101-105

Yoder HS, Tilley CE (1962) Origin of basalt magma: an experimental study of natural and synthetic rock systems. J Petrol 3: 342-532

Zies EZ, Chayes F (1960) Pseudoleucite in a tinguaite from the Bearpaw Mountains, Montana. J Petrol 1: 86-98

Subject Index

Page numbers in *italics* refer to main entries

Minerals and Rocks

Editor Chief: P. J. Wyllie

Editors: W. von Engelhardt, T. Hahn

Volume 10
J. T. Wasson

Meteorites

Classification and Properties

1974. 70 figures. X, 316 pages
ISBN 3-540-06744-2

"The book is a very successful attempt at including all aspects of meteorite studies in one thin volume. It is almost unbelievable that scientific material, originating from so many sciences and of such large extent, can be compressed into one relatively small book. The author achieved this without detriment to clearness, and the book is written definitely and distinclty... (The author) has been one of the leading specialists in meteorite research during in the past decade. He is responsible for a good deal of progress in meteorite science. His survey book on meteorites will be used by experienced specialists as a good reference book and by starting specialists and students as an excellent introduction into meteorite research. Moreover, it will be a handbook for everybody, who wants to classify new or unknown meteorite samples."
Journal of the British Interplanetary Society

Volume 11
W. Smykatz-Kloss

Differential Thermal Analysis

Application and Results in Mineralogy

1974. 82 figures, 36 tables. XIV, 185 pages
ISBN 3-540-06906-2

This work surveys the application of differential thermal analysis (DTA) to problems of crystal physics, crystal chemistry, sedimentary petrography and petrology. The author cites the DTA characteristics of the most important minerals, and explains how the use of improved analytical methods and key diagrams of a new type can aid the systematic determination of minerals by DTA. "This is a compact, critical, and authoritative treatment of DTA, ... References are conveniently listed alphabetically and the subjects are well indexed... The book will be useful to materials scientists and mineralogists and invaluable to specialists in raw materials and thermochemistry."
Ceramic Abstracts

Springer-Verlag
Berlin
Heidelberg
New York

Volume 12
R. G. Coleman

Ophiolites

Ancient Oceanic Lithosphere?

1977. 72 figures, 18 tables. IX, 229 pages
ISBN 3-540-08276-X

This book provides the most comprehensive review now available on ophiolites, which have recently been shown to play an important role in the new plate tectonic hypothesis. Representing fragments of ancient oceanic lithosphere, the origin, tectonic setting and age of ophiolites offer provocative information for reconstruction of ancient plate boundaries and formation of oceanic crust. The data presented here is essential for accurate evaluation of the ancient oceanic lithosphere and will be an indispensable aid for earth scientists and students.

"... This is a useful book, as the text is informative and the diagrams are clear and relevant... The preparation of such a volume is a daunting task for any worker, in view of the subjective stage at which ophiolite research now stands. Dr. Coleman should be congratulated for a very creditable attempt to provide an authoritative text on the subject."
IMM Bulletin

Volume 13
M. S. Paterson

Experimental Rock Deformation – The Brittle Field

1978. 56 figures, 3 tables. XII, 254 pages
ISBN 3-540-08835-0

"... The previous volume in the series, that by R. G. Coleman on ophiolites, was particularly informative and accessible to a non-specialist readership. Paterson maintains this high standard with *Experimental Rock Deformation* despite the, perhaps, more esoteric nature of his subject. He successful breaks through the technological and terminological aura which surrounds "rock squashing"... There is a large reference list of over a thousand items which is comprehensive if not exhaustive. Paterson writes throughout with clarity, economy of style, and total familiarity with his subject. The book is adequately and clearly illustrated and well produced at a reasonable price. The reference list alone will make Paterson's book invaluable to those actively engaged in experimental rock deformation. However, a much wider audience of engineering, mining and structural geologists, and geophysicists will welcome this clear, careful review of an important field..."
Geological Magazine

"... Paterson's monograph is clearly the most authoritative and definitive study of the brittle behaviour of rock available. It will serve us as a useful state-of-the-art summary and clarifier of ideas for reserach workers with some background in the subject, as an in-depth introduction for research workers to whom the subject is relatively new, and as a text- or reference-book for other than beginning students of rock mechanics..."
Int. J. Rock Mech. Min. Sci. & Geomech. Abstr.

A. S. Marfunin

Physics of Minerals and Inorganic Materials

An Introduction

Translated from the Russian by
N. G. Egorova, A. G. Mishchenko

1979. 138 figures, 50 tables. X, 340 pages
ISBN 3-540-08982-9

Solid-state theories and spectroscopy account for the third crucial change within this century in our concept of the basis of mineralogy and the inorganic materials sciences. This book is a revised, updated, and supplemented translation from the Russian edition, providing a complete system of recent theories of solids as they apply to minerals and inorganic materials. Both basic principles and sophisticated new theories are presented with the mineralogist and materials researcher in mind. The book contains extensive references for further study and will be a valuable reference work since each chapter is self-contained.

Contents: Quantum Theory and the Structure of Atoms. – Crystal Field Theory. – Molecular Orbital Theory. – Energy Band Theory and Reflectance Spectra of Minerals. – Spectroscopy and the Chemical Bond. – Optical Absorption Spectra and Nature of Colors of Minerals. – Structure and the Chemical Bond. – Chemical Bond in Some Classes and Groups of Minerals. – References. – Subject Index.

A. S. Marfunin

Spectroscopy, Luminescence and Radiation Centers in Minerals

Translated from the Russian by V. V. Schiffer

1979. 170 figures, 22 tables. X, 352 pages
ISBN 3-540-09070-3

Progress in the research of minerals, lunar specimens, and inorganic materials has been greatly affected in the past decades by the introduction of solid-state spectroscopy methods. This book examines mineralogical applications of nuclear gamma-resonance (Mössbauer) spectroscopy, X-ray and electron spectroscopy, nuclear magnetic resonance and electron paramagnetic resonance. The luminiscence and thermoluminescence of minerals are considered on the basis of crystal field, molecular orbital and energy band theories, as well as EPR data. Models of electron-hole centers for many types of minerals and materials are treated extensively based on molecular orbital theories and EPR studies. Basic principles are outlined and experimental data surveyed in each chapter. The bibliography contains 1,023 references to papers and books published mostly between 1960 and 1970. A unique reference, this book breaks new ground in the earth sciences and presents heretofore largely inaccessible material.

Contents: Mössbauer (Nuclear Gamma-Resonance) Spectroscopy. – X-Ray and X-Ray Electron Spectroscopy. – Electron Paramagnetic Resonance. – Nuclear Magnetic Resonance (NMR) and Nuclear Quadrupole Resonance (NQR). – Luminescence. – Thermoluminescence. – Radiation Electron-Hole Centers (Free Radicals) in Minerals. – References. – Subject Index.

Springer-Verlag Berlin Heidelberg New York